The Theory of Objectification

Semiotic Perspectives on the Teaching and Learning of Mathematics

VOLUME 4

The titles published in this series are listed at *brill.com/sptl*

The Theory of Objectification

A Vygotskian Perspective on Knowing and Becoming in Mathematics Teaching and Learning

By

Luis Radford

BRILL
SENSE

LEIDEN | BOSTON

Cover illustration: Image by Andrew Sanderson

All chapters in this book have undergone peer review.

The Library of Congress Cataloging-in-Publication Data is available online at http://catalog.loc.gov

Typeface for the Latin, Greek, and Cyrillic scripts: "Brill". See and download: brill.com/brill-typeface.

ISSN 2542-8705
ISBN 978-90-04-45964-9 (paperback)
ISBN 978-90-04-45965-6 (hardback)
ISBN 978-90-04-45966-3 (e-book)

This book is printed on acid-free paper and produced in a sustainable manner.

A Dora, por su infinita paciencia

Contents

Preface XI
List of Figures and Tables XIV

Introduction: The Ascent from the Abstract to the Concrete 1

1 **Theories in Mathematics Education** 2
1 Outline 2
2 A Classroom Episode 2
3 Research Questions 6
4 Method 6
5 Theoretical Principles 7
6 Piaget's Genetic Epistemology 7
7 From Method to Methodology 8
8 Mathematics Education Theories: Two Short Examples 9
9 The Theory of Objectification 14

2 **An Overview of the Theory of Objectification** 17
1 Outline 17
2 Introduction 17
3 Theoretical Underpinnings of the Theory of Objectification 20
4 Summing up and Looking Ahead 34

3 **Knowledge and Knowing** 38
1 Outline 38
2 Knowledge 38
3 Knowing 49
4 The Piggy Bank Example 52
5 The Dialectic between Knowledge and Knowing 53
6 Mathematics as an Entity at the Same Time *Ideal, Sensible, and Material* 56
7 Synthesis 59

4 **Learning** 63
1 Outline 63
2 Learning as Participation in Social Practice 63
3 Internalisation 65
4 Processes of Objectification 71
5 Some Meanings of Objectification 76
6 Processes of Objectification: A Processual Definition 78

7 Learning as Objectification 79
8 Consciousness 79
9 Teaching-Learning Activity 80
10 Processes of Subjectification 82
11 Synthesis 83

5 **Processes of Objectification** 86
1 Outline 86
2 The Investigation of Processes of Objectification 86
3 Teaching-Learning Activity 88
4 An Example of Investigation of Processes of Objectification 92
5 Semiotic Means of Objectification 100
6 Semiotic Nodes 101
7 Semiotic Contraction 102
8 Concept 102
9 Synthesis 104

6 **Embodiment** 108
1 Outline 108
2 Introduction 108
3 The Intertwining of the Senses and Culture 110
4 Perception 114
5 A Classroom Example 116
6 The Poetry of Objectification 125
7 Counting the Unseen 126
8 Synthesis 128

7 **Task Design: Or Configuring Teaching-Learning Activities** 132
1 Outline 132
2 General Considerations 132
3 The Motion of Tina, John, and the Dog 135
4 Synthesis 149

8 **The Cultural Nature of Mathematical Thinking** 156
1 Outline 156
2 Introduction 157
3 Boas's Relativist Conception of Culture 159
4 The Anthropological Venerable Conflict 161
5 A Dialectical Materialist View of Culture 163
6 Greek Mathematical Thinking Revisited 167
7 Synthesis 177

9 **Processes of Subjectification** 183
1 Outline 183
2 The Question of the Subject 183
3 Semiotic Systems of Cultural Signification 185
4 Being, Becoming, and Subjectivity 188
5 Solving Equations in a Grade 3 Classroom 192
6 Synthesis 201

10 **Ethics** 205
1 Outline 205
2 The Ineludible Presence of Ethics in the Mathematics Classroom 206
3 Kant 208
4 Hobbes 210
5 Lévinas's Ethics 211
6 The Indispensable Task of (Mathematics) Education 216
7 Towards a Communitarian Ethics 219
8 Synthesis 226

References 231
Index 256

Preface

This book presents a systematic account of a sociocultural theory: the theory of objectification.

The theory of objectification seeks to overcome the long-standing Western tradition that conceives of learning as the result of subjective experience. This notion of learning can be traced back to the 18th century where the child was imagined as a naturalistic being already endowed with her own potentialities. In this line of thought, what is required for the child to learn is merely a conducive environment. In our times, this conception of learning has been epitomised by child-centred pedagogies and, more recently, by the various forms of constructivism (e.g., psychological, social, radical). In these educational approaches learning remains, however, blind to considerations of a societal, cultural, historical, and political nature. As a result, learning is abstracted from the antagonistic and conflicting forces of society and ends up portrayed as a sanitised process where what remains is merely reified subjective experience. Moving away from the naturalistic and subjectivist premises of the child-centred pedagogy and constructivism, the theory of objectification is an invitation to envision learning as a genuine collective cultural-historical process.

The perspective on learning to be explored in this book depends, of course, on how the collective is conceptualised. It could be argued, indeed, that, one way or the other, learning involves some sort of collective—even in the pedagogical model of direct or transmissive teaching. The collective that the theory of objectification brings to the fore does not refer here to an assemblage of individuals—Piagetian-like monads in interaction—idiosyncratically negotiating their own meanings. It refers rather to individuals who, headed to the satisfaction of communal needs, work *together* (most of the time with tensions, conflicts, and contradictions); individuals—teachers, students—who, in working together, *learn together* and co-produce themselves against the always contested background of culture and history. What glues the individuals of a collective together is cultural-historical activity.

The theory of objectification finds its inspiration in Hegel's and Marx's philosophies, the seminal work of L. S. Vygotsky and his collaborators, and Freire's concept of education. Its goal is twofold:

- First, to offer a precise theoretical conception of learning as a genuine collective agentic cultural-historical process.
- Second, to explore the practical pedagogical conditions that make genuine collective learning possible.

In this sense, the theory of objectification is more than a lens through which to look at, and interpret, learning phenomena. It is also an invitation to *trans-*

form pedagogical practices and to *transform* the classroom into a site where students can encounter cultural knowledges and voices in deep conceptual ways while at the same time making the experience of collective life, solidarity, plurality, and inclusivity.

A prolegomenon to such a task is an entire reconceptualisation of teachers and students. Blurring the customary borders that, in other approaches, separate teachers and students, the theory of objectification moves beyond the concept of the teacher as a patriarch of knowledge—a sort of coach helping the students grasp curricular contents through scaffolding or other strategies. The theory of objectification articulates the idea of teachers and students as beings in flux, as unfinished and continuously evolving projects of life, in search of themselves, co-producing themselves every day as they engage ethically in a *same* endeavour where they suffer, struggle, and find enjoyment and fulfillment together.

This book grew out of discussions, conferences presentations, and seminars to researchers and undergraduate and graduate students conducted in various countries. It capitalises on the cultural diversity of the audiences. It also capitalises on the sustained work with prospective teachers and in-service teachers and their students in Canada over a span of more than 20 years. It is, indeed, in the arena of concrete classrooms that the questions that led to the theory were formulated. The polemics that the aforementioned discussions and presentations engendered at home and abroad are unavoidably a part of the concepts that came to form the core of the theory of objectification. These concepts are bearers of dissident voices and perspectives.

Some of the ideas presented in the book are revisions, refinements, and expansions of ideas that appeared in previous papers. Chapters 3, 4, and 5 have their origin in my paper "Three key concepts of the theory of objectification" published in the *Journal of Research in Mathematics Education*, 2(1), 7–44, 2013, following a one-week seminar organised by Oh Nam Kwon in South Korea. Chapter 6 is an expanded and modified version of "The eye as a theoretician," published in *For the Learning of Mathematics*, 30(2), 2–7, 2010[1] and prepared first for a seminar organised by Bernadete Morey at the Universidade Federal do Rio Grande do Norte, Brazil. The last chapter of this book includes passages from an in-press paper that will appear in the book "Ética: entre educación y filosofía" to be published by Ediciones Uniandes, Colombia, after a seminar organised by Maritza Silva Acuña at the Universidad Católica Silva Henriquez in Santiago de Chile and a paper recently published in a Special Issue of the *Revista Colombiana de Matemática Educativa*, edited by Rodolfo Vergel and Isaias Miranda. I am grateful to the publishers for the permission to use those materials.

I hope that this book will make a contribution to education in general and mathematics education in particular. I wish to thank Sandra Siren and Andrew Sanderson for their help in the preparation of the book. I also want to acknowledge the continued support of the Social Sciences and Humanities Research Council of Canada/Le conseil de recherches en sciences humaines du Canada (SSHRC/CRSH), and the Laurentian University Research Fund.

Note

1 See https://flm-journal.org

Figures and Tables

Figures

1 Left: Krysta pointing to the bingo chips in front of Week 5. Right: Albert pointing to the blue bingo chip. 3
2 The doubling strategy, with the corresponding removal of one bingo chip. 4
3 The students and the teacher discussing the strategy to find the number of bingo chips in Week 10. 5
4 A theory as a dynamic unit made up of three components (P, M, Q). 9
5 Vygotsky's famous triangle. External signs and other components of material and spiritual culture, X, alter the human mind's natural history (Vygostky & Luria, 1994, p. 144). 40
6 Yo, the middle chimpanzee, breaks a nut while the young chimpanzees attentively follow the process (from Matsuzawa, Biro, Humle, Inoue-Nakamura, Tonooka, & Yamakoshi, 2001, p. 570). 42
7 The first terms of a sequence that students in Grade 2 investigated in a lesson. 44
8 The first triangular, square, and pentagonal numbers. 45
9 Through activity, knowledge (*K*) acquires a sensible content, and is materialised as knowing (*k*). New systematisations occur, leading to an expansion of knowledge. The resulting new knowledge (*K'*) leads through activity to new knowing (*k'*), etc. 54
10 Grade 6 students materialising a cultural-historical algebraic way of thinking (knowledge) through a classroom activity. The materialisation consists of reflections and actions on the given equation. 55
11 Frequently reported imagery of the term 100 for trained eyes. 74
12 Terms 5 and 6 drawn by a Grade 2 student (7–8 years old). 75
13 A student investigating the sequence. Thinking aloud, he points out the bottom row of Term 2 (left), then the top row of Term 3 (right). 75
14 Modelisation of the learning process. The students put the tokens inside the goblets. 87
15 A model of an arithmetic sequence with bingo chips of two colours. 87
16 On the left, the structure Φ of the activity. On the right, the activity Θ itself; Θ is a function of the didactic project and other variables. 90
17 The activity Θ and some of its "moments". 91
18 The students' strategy for calculating the savings in Week 10. 93
19 Mrs. Giroux and Albert engaged in joint labour. 95

20 Coordination of gestures on the blackboard with what Albert says from his desk. 97
21 A sequence in an assessment that took place more than a week later. 98
22 The first four terms of a sequence investigated in a Grade 2 class. 117
23 The term drawn by Monique. Is this Term 8 of the sequence? 117
24 Left: the moment in which a student (James) is drawing Term 6. Middle: James's Terms 5 (top) and 6 (bottom). Right: Term 8 according to another student (Sandra). 118
25 The teacher suggesting a perceptual manner to see the terms of the sequence. 120
26 The teacher pointing to the empty space where one would expect to find Term 8. 121
27 The teacher and the students explore the bottom row of the terms. 124
28 (a) The teacher points to the last square in the top row of Monique's term while Sandra says "8." (b) Astonished by the result, Sandra stays still and silently looks at the teacher for 2.5 seconds. 125
29 The teacher and the students counting rhythmically say (Picture 1) "Term 5," (Picture 2) "5 on the bottom," (Picture 3) "5 on top," (Picture 4) "plus 1". 127
30 The design of the problems based on a conceptual and contextual unity and an increasing conceptual complexity. 134
31 The general structure of the component Φ associated with the teaching-learning activity. 137
32 Left: The time-distance graph representing Tina's motion. Right: The graph representing John's motion. 138
33 Left: the students measuring distances on the floor. Right: the students practising Tina's walk. 140
34 The Calculator Based Ranger® (left) and the graphing calculator (right) used by the teacher. 140
35 Problem 1 and its three actions. 141
36 Table of values that relates the distance travelled by the dog at certain values of the variable time. 143
37 Problem 2 consisted of 2 actions and 2 questions. 145
38 Problem 3 consisted of 3 actions and 2 questions. 147
39 A small group's answer to the last question. 148
40 Problem 4 consisted of 1 action and 3 questions. 149
41 The design of the task based on a conceptual and contextual unity and an increasing conceptual complexity. 149
42 Some lines from the stele identified as *IG*, I² 324 (from Lang, 1964, p. 146). 171
43 The diagram accompanying the proof of Euclid's Book I, Prop. 34 (from Heath, 1908, p. 324). 174

44 Discussing equations written through concrete artefacts. 193
45 Pictures 1 and 2 show Team A and the corresponding team (Team B) working independently to produce a mathematical text. Picture 3 shows Team B critically examining the other team's text. The other team is doing the same. Picture 4 shows that the members of the two corresponding teams meet to discuss their texts; from left to right, Sara, Sandra, Carl, Elisa, and Christine. 195
46 Top row: equation and solution of Team B. Bottom row: equation and solution of Team A. 197
47 Top: the teacher offering a different interpretation concerning the place of the mathematical question. Bottom: Carl refuting the teacher's interpretation. 200
48 The unfolding ethical relationship between the teacher and the students depends crucially on the body. 224

Tables

1 A synoptic panorama of the theory of objectification. 35
2 Total time of two famous recordings of Beethoven's 7th Symphony. 58

INTRODUCTION

The Ascent from the Abstract to the Concrete

This book is about a theory of teaching and learning: the theory of objectification. The theory is inspired by contemporary dialectical materialism. One of the characteristics of dialectical materialism is a principle according to which knowing proceeds from the *abstract* to the *concrete*. Knowledge appears first in a *general* or *abstract* manner. As our journey unfolds, knowledge acquires specific or *concrete* content. In the terminology of dialectical materialism, such a journey is called the *ascent from the abstract to the concrete*.

This book is an example of this ascent. Our journey starts from the *general*: What is a theory in mathematics education? Then, in the following chapters, the concepts of the theory of objectification appear, showing their concrete content and interrelated components. As we travel through the book, we encounter the concepts of knowledge and knowing, being and becoming, learning, the meaning of culture and embodiment, and, at the end, we reach the most concrete aspect of human cultural-historical activity, namely ethics.

 | DOI: 10.1163/9789004459663_001

CHAPTER 1

Theories in Mathematics Education

1 Outline

In this chapter, my starting point is that a theory in mathematics education is something like a *system*. It is always changing and is made up of interrelated components. I want to suggest that a theory includes three interrelated components: (1) an organised ensemble of theoretical principles, (2) a methodology, and (3) some paradigmatic research questions. I will develop this idea in this chapter. My discussion starts with a Grade 4 classroom episode, which I use to progressively introduce the aforementioned components. Then, I briefly discuss two examples of theories in mathematics education—constructivism and the theory of didactic situations—and finish the chapter with a short reference to the theory of objectification.

2 A Classroom Episode

Before starting her Grade 4 class, in front of all the students, the teacher—Mrs. Giroux—asked Marc to tell the class something about his recent birthday. Marc mentioned some of the gifts he had received. Among the gifts was a piggy bank of which Marc seemed particularly fond.

It happened that Mrs. Giroux and her Grade 4 class of 9–10-year-old students were to soon start dealing with algebraic pattern generalisation. Taking advantage of Marc's gift, Mrs. Giroux came up with a story through which the students could start investigating a covariational approach to produce a formula (not necessarily using alphanumeric symbolism). In terms of the curriculum, the formula should allow the students to calculate the value of remote terms in a simple sequence. A few days later the students were working in small groups on a problem based on the following story:

> For his birthday, Marc receives a piggy bank with one dollar. He saves two dollars each week. At the end of the first week he has three dollars, at the end of the second week he has five dollars, and so on.

The teacher provided the students with bingo chips of two colours (blue and red) and numbered plastic goblets intended to represent Week 1, Week 2,

 | DOI: 10.1163/9789004459663_002

etc., and invited the students to model the saving process until Week 5. Then, drawing on the model, the teacher invited the students to find the amount of money saved at the end of Weeks 10, 15, and 25.

Here is an excerpt from the discussion of a three-student group—Albert to the right (partially visible in the pictures below), Krysta in the middle, and Manuel to the left.

14. Krysta: So, we should do ... That (*see Figure 1, left*) times two. So, 11 ...
15. Albert: 11 plus 11 ... 22.
16. Krysta: 22.
17. Albert: (*He laughs.*)
18. Krysta: At the end of ... okay, at the end ...
19. Albert: Well, wait ... No. It would be 11 plus 10 because ...
20. Krysta: (*Pointing to Week 5*) 5.
21. Albert: (*Pointing to the blue bingo chip*) We always start with the ... [blue chip] (*see Figure 1, right*).

FIGURE 1 Left: Krysta pointing to the bingo chips in front of Week 5. Right: Albert pointing to the blue bingo chip

As we can see, to answer the question about Week 10, the students resort to a doubling strategy. They add the amount of dollars saved until Week 5, double it, and proceed to remove one from that amount, which corresponds to a blue bingo chip, since there is always one and only one blue bingo chip in a week (see Figure 2).

The reason is not perfectly articulated by Albert in lines 19 and 21, but his clear pointing gesture helps fill the verbal gap in the intended explanation.

The doubling strategy can be applied to other cases, as the students noted. Hence, to find the amount of saved money at the end of, say Week 25, we observed some groups start from Week 5; by doubling, they get the amount of saved money in Week 10. They double again and add the amount of Week 5,

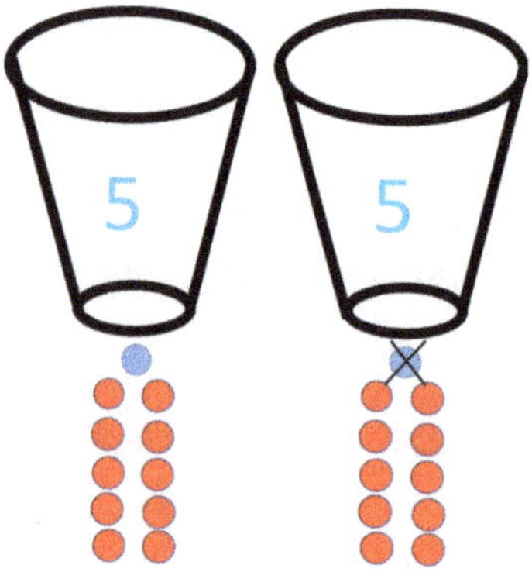

FIGURE 2
The doubling strategy, with the corresponding removal of one bingo chip

without forgetting to remove the extra blue chips that were added in the process. The strategy works just fine but is cumbersome to determine the amount of savings in "remote" weeks such as Week 78 or 103.

But let us continue with the Piggy Bank example. When the teacher (Mrs. Giroux) went to see Krysta and her teammates' work, she realised that the students were resorting to the doubling strategy. The discussion went as follows:

22. Mrs. Giroux: What I find interesting here is that [in your model] you have bingo chips of two colours. What does it mean?
23. Krysta: Because blue was what he [Marc] already had.
24. Manuel: Yeah, because it [the story problem] says that the piggy bank had 1 dollar.
25. Albert: The piggy bank had 1 dollar, so those (*He points in sequence to all the blue bingo chips from Weeks 1 to 5*) are all the 1 dollars that he had already (*Now he points to the red bingo chips from Weeks 1 to 5*) added to 2, 4, 6, 8, 10.
26. Mrs. Giroux: Okay, okay. What would happen if it was Week 10?
27. Albert: Well, (*Pointing to Week 5*) we added all this again (*He makes a sweeping gesture; see Figure 3, Pictures 1 and 2*), because we know that 5 + 5 = 10, so …
28. Krysta: (*Interrupting*) Plus … we added … We added all this (*She points to the red bingo chips in Week 5; see Figure 3, Picture 3*), not the blue (*She points to the blue bingo chip …*).
29. Mrs. Giroux: (*Trying to make noticeable to the students the co-variational structure*) What do you remark about Week 5 (*She shows the glass corresponding to Week 5*) and (*Pointing to the red bingo chips; see Figure 3, Picture 4*) the number of bingo chips? (*Making the same actions*) The fourth week and the number of bingo chips?
30. Albert: It's always twice …

31. Mrs. Giroux: (*Repeating*) It's always twice.
32. Krysta: It's the double of what you ... No! (*She watches the artefacts intensely for a while*) I am confused!
33. Albert: Yeah! It's twice, look! (*Counting the red chips*) 1 + 1, 2; 2 + 2, 4; 3 + 3, 6; 4 + 4.
34. Krysta: (*Interrupting*) 8.
35. Albert: (*At the same time*) 8.
36. Krysta: 5 + 5, 10.
37. Albert: (*Pointing to the bingo chips in Week 5*) 5 + 5, 10.
38. Krysta: Cool. It's twice the week ...
39. Mrs. Giroux: So, if the reds are twice [the number of the week], what happens to the [blue] bingo chip? (*She points to the blue bingo chip in Week 5*)
40. Krysta: Plus 1.

From lines 22 to 28 the students explain their strategy to the teacher. The strategy is not identified with a label (such as the doubling strategy or something else); rather, the strategy is explained in action, through words and gestures.

In line 29 something important happens. Indeed, in line 29, the teacher tries to introduce what seems to be a new approach to perceiving things: She says, "What do you remark about Week 5 and the number of bingo chips? The fourth

FIGURE 3 The students and the teacher discussing the strategy to find the number of bingo chips in Week 10

week and the number of bingo chips?" At the end of the passage, the students seem to start noticing a co-variational relationship between the number of the week and the red and blue chips.

3 Research Questions

Through its research questions, a theory addresses specific *problems* that it seeks to investigate. For example, a theory interested in the problem of classroom *interaction* would ask questions about how the teacher-students interaction occurs and perhaps questions about how it evolves during the year or over the years. A theory interested in the problems of teaching and learning would ask questions such as: *What* are the students learning? *How* are they learning it? Such a theory would also ask questions about the teacher. For instance, in reference to the classroom episode reported above, the theory could ask: What is the nature of the teacher intervention? Is her intervention appropriate?

4 Method

To answer these and other similar questions, a theory needs to produce some *facts.* Facts are indeed required in order to *articulate accounts* of the phenomenon under scrutiny in a *convincing manner.* So, on what parts of the Piggy Bank example should we focus? Which *parts* of the episode should we keep as *facts* of the students' interaction (in the case of the first theory) or the students' learning (in the case of the second theory)?

The previous comment makes it already apparent that the selection of facts and their eventual interrelation are not produced at random. They are produced *systematically*, through some *method.*

A method brings some facts forward and leaves in the background that which the method deems "unimportant"—*noise.* So, if we want to answer some of the aforementioned questions (e.g., "How are the students learning in the Piggy Bank example?" assuming that they are learning *something*), should we focus on the students' actions? Should we focus on the students' language? Should we attend to the goblets and bingo chips and the teacher's and students' gestures, body positions, and prosody (e.g., verbal intonations)? What should we attend to exactly? There are many possibilities. They depend on something that goes *beyond* the realm of facts—the realm of what some philosophers and linguists call *Weltanschauung* (see Underhill, 2009); that is, a *worldview* (our implicit and explicit conceptions about how the world *is*). It is, indeed, against the background of our worldview and *its theoretical principles* that facts are *interpreted.*

5 Theoretical Principles

Take the example of the ancient Greeks. Following Aristotle, the ancient Greeks believed that facts were bearers or embodiments of universal properties. This is an example of a *theoretical principle.* It gives facts a precise place in the theory. And it also endows the method with a precise function. In any theory, the theoretical principles and the method that produces facts and interpretations must fit mutually. We see this fit very well in the case of the ancient Greeks. For them, the function of a method was to help one make sense of concrete or ideal things *already out there* by looking at them attentively. A method for them was a kind of *contemplative* process. Classifications, like the botanical ones carried out by Aristotle, were methods to sort out things and to understand them. Thus, it follows, as Heidegger (1977) suggested, that for the ancient Greeks *theōrein* or *theory* was a form of seeing, or looking at something attentively, so to make it reveal itself through the spectacle of its appearance.

It is interesting to note that, in the ancient Greek worldview, the observed objects were not forced to appear. We have to wait until the late Middle Ages and early Renaissance to find the idea that we can *force* the objects under scrutiny to appear. That was the role of the *scientific experiment* (Dear, 1995; Shapin, 1994). The scientific experiment brought with it a change in the conception of facts. Indeed, since the emergence of empiricism, and in particular in the early 17th century, under the influence of Francis Bacon (1906), facts became understood as theory-free particulars. As Mary Poovey (1998, p. xviii) notes in her book, *A History of the Modern Fact*, some scientists argued that "one could gather data that were completely free of any theoretical component." The theoretical principle here is that facts *generate* knowledge about the world through an inductive process.

Before we continue our discussion on theories, and to illustrate what we have seen so far, I want to consider Piaget's theoretical approach.

6 Piaget's Genetic Epistemology

Three of the theoretical principles of Piaget's genetic epistemology are the following:

1. Universal cognitive mechanisms underpin the child's construction and development of cognitive operations and structures.

2. These cognitive structures are describable in terms of logico-mathematical structures.
3. The construction of the child's cognitive structures is the result of the child's own deeds.[1]

From these theoretical principles Piaget devised a method based on clinical interviews. The clinical interviews offered Piaget an ideal space for experimentation and observation.

In the interviews, an adult presented meticulously prepared problems and/or question(s) to a child. The problems or question(s) were engineered to allow Piaget to produce *facts* and interpretations. From those facts and their interpretations Piaget was able to articulate narratives to answer his developmental questions.

The substance or material from where Piaget's facts were produced was the child's actions (e.g., in handling some concrete objects) and utterances. They constituted the raw data from where Piaget and his collaborators tried hard to find traces of logical thinking. As Gardner pointed out,

> Piaget examines a child's protocol and picks out the significant underlying propositions (which he can then order in the logical parlance of p's and q's); the mental action reflected in the protocol is a series of operations performed on the propositions. The individual has reached formal operations when he [sic] can systematically and exhaustively explore the relations between propositions describing a phenomenon. (Gardner, 1970, p. 359)

Piaget's method allows him to discern *facts* among what the child says and does. By being entangled with a worldview and theoretical principles about the human mind, the method allows Piaget to *interpret* the facts and to produce accounts or narratives to prove something about the child's intellect.

As in the case of the ancient Greeks, we see a proper and apt adequation between: (a) method and its production and interpretation of facts, and (b) theoretical principles. Methods and theoretical principles are aligned in such a way that accounts can be convincingly articulated to answer the theory's research questions.

7 From Method to Methodology

There is something that needs to be said about the nature of the theoretical principles of a theory, as it has important consequences in the understanding of theories in general: it is impossible to make *explicit* all the principles of a theory. It is only in formal theories (e.g., axiomatic set theory) that a list of principles (axioms) is possible. Theories in mathematics education (and in the social sciences more generally) are *not* formal theories. Some of their principles (e.g., those that are deemed to be more important) can be made explicit; other principles, however, run in the background. As a result, the fit between method and principles (a fit that makes it possible to interpret facts) needs to be seen as something not necessarily permanent. There is always a tension (a dialectical tension) between the theoretical principles and the method of a theory. Furthermore, as the theory produces results, theoretical principles and methods may be affected. This is why it is better to think of a theory as a system in motion and transformation. One of the consequences of such a view of theories is that methods, notwithstanding the empiricists, cannot be reduced to a series of steps to be rigorously followed. A method is something much more complex. This is why, for Vygotsky (1997a, p. 27), "Finding a method is one of the most important tasks of the researcher." For Vygotsky, the main characteristic of a method is to be inquisitional and reflective. In Vygotsky's view, a method is a truly *philosophical practice*. To emphasise this non-instrumentalist conception of method, in what follows, I will use the term *methodology* instead of *method*.

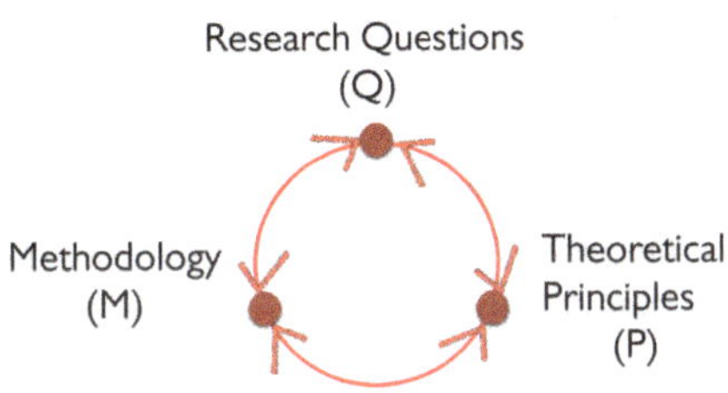

FIGURE 4
A theory as a dynamic unit made up of three components (P, M, Q)

I can summarise the previous discussion by saying that a methodology, M, can make sense only through its interrelationship with a set of theoretical principles, P. These principles convey implicit and explicit features of the underpinning worldview, and the research questions, Q, which the theory seeks to answer. Elsewhere (Radford, 2008a, 2017a), referring to the research field of Mathematics Education, I have termed *theory* the triplet (P, M, Q). The arrows in Figure 4 indicate that there is a dialectical relationship between them. In their interrelationship, each one of them alters the others, making the theory a dynamic system.

8 Mathematics Education Theories: Two Short Examples

I limit myself here to a brief discussion of two major theories in mathematics education: constructivism (Cobb, 1988; Cobb & Yackel, 1996) and the theory of didactic situations (Brousseau, 2002).

8.1 *Constructivism*

Ernst von Glasersfeld formulated in a clear and explicit manner the principles of (radical) constructivism. The principles are:

p1: Knowledge is not passively received but built up by the cognising subject.

p2: The function of cognition is adaptive and serves the organisation of the experiential world, not the discovery of ontological reality (see von Glasersfeld, 1995, p. 18).

The true educational novelty of constructivism does not rest on the first principle. The first principle was the main axis around which the progressive pedagogues of the so-called educational reform of the early 20th century built their pedagogical program (Darling & Nordenbo, 2002; Röhrs & Lenhart, 1995; Rugg & Shumaker, 1969)—although they did not articulate it as clearly as von Glasersfeld did. This principle also constituted a cornerstone in Piaget's (1970) genetic epistemology (see principle 3 in Section 1.5). The true novelty of constructivism rests rather, as von Glasersfeld claims, on the epistemic and ontological stances conveyed by the second principle and its concomitant concept of knowledge. Without necessarily denying the existence of a pre-existent reality, and in a move coherent with Kant's (2003) theory of knowledge, constructivism does not claim that the knowledge constructed by the cognising subject corresponds to such a reality. Its epistemology rests precisely on the denial of the possibility of any certain knowledge of reality (Ernest, 1991). There is a third principle that, although not mentioned explicitly, operates ubiquitously in constructivist classroom practice. The third principle is not exactly a principle of an epistemological or ontological nature. The third principle provides us with an idea of how knowledge construction should occur in instructional settings and is hence of a pedagogical nature. The principle can be stated as follows:

p3: The cognising subject not only constructs her own knowledge but she should do it in an autonomous way.

Intellectual autonomy was in fact part of two of the general goals identified by constructivism from the outset: "teaching by imposition is incompatible with two general goals of mathematics instruction that follow from constructivism, the construction of increasingly powerful conceptual structures and the development of intellectual autonomy" (Cobb, 1988, p. 99).

These principles provide the basis to conceptualise learning: learning is equated with doing; that is, to the student's actions, which, transformed into schemas, come to constitute *viable* knowledge (von Glasersfeld, 1995) or *taken-as-shared* knowledge (Cobb, Yackel, & Wood, 1992). Furthermore, these principles are in interrelation with the methodological production and interpretation of facts that count as admissible evidence of learning. Constructivists explore the students' actions through what they do or say resorting to a specific teaching experiment design that can last long periods of time—e.g., one year. In Cobb, Stephen, McClain, and Gravemeijer (2001) the constructivist methodological approach involving sequences of instructional activities is explained, as well as the interpretive framework that they use to organise their analyses of classroom events. Since constructivists make a clear division between the social on the one hand, and the individual on the other hand, their analyses work on the basis of a coordination of these two dimensions: they coordinate a social perspective on classroom activities with a psychological perspective that tries to grasp the reasoning of the students.

These principles and their concomitant methodology may allow us to assert, from a constructivist viewpoint, whether or not Krysta, for instance, is learning in the Piggy Bank classroom example above. They may also allow us to make assertions about the appropriateness of the teacher's involvement.

Let me now turn briefly to the theory of didactic situations (TDS).

8.2 *The Theory of Didactic Situations*

The TDS seeks to offer a model, inspired by the mathematical theory of games, to investigate in a scientific way, the problems related to the teaching of mathematics and the means by which to enhance it.

In the beginning, the term "situation" referred to the student's environment as handled by the teacher for whom it appears as a tool in the process of teaching. Later, the situation was enlarged in order to include the teacher herself and even the educational system as a whole (Brousseau, 2003). We see here an example of the evolution or transformation of a theory.

The TDS works on a set of principles, among them the following epistemological ones:

- p_1: Knowledge results as the "optimal" solution to a certain situation or problem.
- p_2: Learning is—in accordance with Piaget's genetic epistemology—a form of cognitive adaptation.

The first principle is connected to a third principle that can be stated as follows:

p_3: For every piece of mathematical knowledge there is a family of situations to give it an appropriate meaning.

This family is called a *fundamental situation*. For Brousseau (2002, p. 30), the search for fundamental situations and their insertion into the more general classroom project of teaching and learning require at least two elements: a good epistemological theory, which would reveal the deepness of mathematical knowledge and inform the teaching process, and a good didactic engineering, which would be oriented to the design of situations and problems to be solved by the students.

The "optimal" solution stated in principle p_1 refers to the mathematical one, as recognised by professional mathematicians, the school, and the curriculum.

A fourth principle specifies further the concept of learning in the TDS expressed in principle p_2. Indeed, the general principle p_2 tells us that learning is of an adaptive nature; it consists of the students' adaptations to a milieu, but it does not say anything about the social-interactional conditions to be fulfilled for it to occur. Principle four is not stated explicitly, but operates in practice within the TDS and provides the foundation for other key concepts of the theory (e.g., the so-called adidactic situation, the Topaz Effect, and a practical distinction between genuine and apparent learning). Principle four is the following:

p_4: The student's autonomy is a necessary condition for the genuine learning of mathematics.

As we can see, principle four has to do with the relationship between teachers and students. This principle gives an impeccable theoretical consistency to the TDS—although, as Brousseau himself recognised, it leads to some practical paradoxes. Thus, according to principle 4, if the process of learning is not accomplished autonomously vis-à-vis the teacher, learning, according to the TDS, has not happened. For "if the student produces her answer without having had herself to make the choices which characterise suitable knowledge and which differentiate this knowledge from insufficient knowledge, the evidence [of learning] becomes misleading" (2002, p. 41). In other words, if "the teacher teaches her [the student] the result, she [the student] does not establish it herself and therefore does not learn mathematics" (Brousseau, 2002, pp. 41–42). The paradoxical juncture lies on the limits imposed to the agentive space of teachers and students: the teachers say but really cannot say (for a discussion, see *The Paradoxes of Learning* in (Radford, 2018a)). Here teachers and students navigate the paradoxical juncture through a contractual stance—a *didactic contract* (I come to this idea in Chapter 10).

As in the case of constructivism, the TDS's principles and methodology provide the basis to make assertions about the students' learning. The assertions will take into consideration the meaning of what doing and learning mathematics mean within the TDS. As Brousseau explains, doing and learning mathematics "must at times be similar to [the] scientific activity [of the mathematicians]" (2002, p. 22). As a result, "A faithful reproduction of a scientific activity by the student would require that she produce, formulate, prove, and construct models, languages, concepts and theories" (Brousseau, 2002, p. 22).

The didactic design of activities should allow the students to investigate, produce, and formulate problem-solving methods:

> The teacher must therefore simulate in her class a scientific microsociety if she wants the use of knowledge to be an economical way of asking good questions and settling disputes, and if she wants language to be a tool for mastering situations of formulation and mathematical proofs to be a means of convincing classmates. (Brousseau, 2002, p. 23)

It is against the background of those ideas and principles that the researcher can assert whether or not the students are learning in our introductory Piggy Bank example (or any other teaching and learning situation for that matter). The TDS's principles and methodology provide room also to make claims about the pertinence of the teacher's intervention.

Constructivism and the TDS present some similarities (e.g., both theories emphasise the importance of the student's autonomy vis-à-vis the teacher). However, they differ in other aspects. For instance, they do not have the same conception of learning, in part because of differences in their concepts of knowledge. In the TDS, knowledge has a social-institutional nature, as opposed to the subjective meaning it has in constructivism. As a result, they may end up with different claims or reasons about the students' learning in the Piggy Bank example or any other example.

There is an important point where both theories may agree: they may agree with the claim that the teacher's *intervention* has a palpable effect on the students' conceptualisation. The teacher's intervention *directs* the students' attention to a co-variational form of thinking about the generalisation problem. In line 29 the teacher tries indeed to make noticeable to the students a co-variational structure when she says: "What do you remark about Week 5 and the number of bingo chips? The fourth week and the number of bingo chips?" Since the co-variational form of thinking does not come from the students, the teacher may be perceived as impeding the students from making their own choices. Not only does the teacher's intervention (as seen through the lenses of constructivism) come to interfere with the "intellectual autonomy" of the students (Cobb, 1988, p. 99), but (as seen through the lenses of the TDS) may also produce misleading evidence of learning.

9 The Theory of Objectification

The Piggy Bank example introduced at the beginning of this chapter provided raw material to remind us that the interpretations and claims that can be made about classroom phenomena depend on the theory to which we resort to analyse them (see also Bikner-Ahsbahs & Prediger, 2014; Niss, 1999; Scheiner, 2020; Scheiner & Pinto, 2019).

I have made a very short incursion into two of the major theories in mathematics education: constructivism and the TDS. Although too brief to do justice to the complexity behind each one of these theories, the incursion into them does allow us to see that theories convey specific conceptions of knowledge that have a great impact on the manner in which learning and teaching are understood. But educational theories also convey ideas about teachers and students. We have seen, for instance, the importance that both constructivism and the TDS give to autonomy. Such a stance draws, of course, on more general ideas of individual and human nature. I will have time to dwell on this point later on in this book.

To make these points, I could certainly have chosen other theories as well. However, that would have moved me away from the focus of this book, which is not about theories in mathematics education. I can only say that perhaps my choice is biographical in nature. I live in Canada where constructivism has had a tremendous influence. And it was in France, during my PhD studies, that I was influenced by the TDS. It is hence my immersion in the French intellectual tradition as a student, on one hand, and my professional activities as a mathematics educator in Canada, on the other hand, that may explain my choice. This biographical reason may also explain the tremendous impact that these theories have had on the theory that is the subject of this book: the theory of objectification.[2]

Yet, greatly influenced as it is by constructivism and the TDS, the theory of objectification departs from them. In what sense does the theory of objectification depart from those theories? There is a short, an intermediate, and a long answer to that question. The long answer can only come after reading this book. To avoid disappointments, in the next chapter I offer an intermediate answer. Indeed, in that chapter I present an overview of the theory of objectification and its place in the landscape of theories in which it belongs: the contemporary mathematics education sociocultural theories. But since the reader may still find this course of action disappointing, let me offer in the rest of this chapter the short answer.

Constructivism and the TDS draw on theoretical ideas that go back to the philosophies of the Enlightenment, and to Kant's philosophy in particular—maybe not directly; maybe as Kant's ideas were refined and re-stated by Piaget (and von Glasersfeld, in the case of constructivism). The theory of objectification draws on a different philosophical tradition, namely Hegel's (1991) philosophy and the ensuing dialectical materialism (e.g., Artinian, 2017; Fedoseyev, Oizerman, Melukhin, Ilyenkov, Lektorsky, et al., 1977; Fischbach, 2014; Ilyenkov, 1977; Lefebvre, 2009; Levant & Oittinen, 2014). The theory of objectification's main sources are not Kant and Piaget, but Hegel, Vygotsky, Leont'ev, Luria, and other psychologists such as Rubinstein (1983). In practice, what it means is that the concepts of knowledge, teaching, and learning, and the very concept of students and teachers, are not the same as those that have come down to us from the Enlightened tradition to inform constructivism and the TDS.

There is still another short answer that may help make sense of the differences I am trying to pinpoint. The theory of objectification is embedded in an educational project that is not the same as the one adopted by constructivism or the one followed by the TDS. As we saw above, constructivism identifies two general goals of mathematics education: "the construction of increasingly powerful conceptual structures and the development of intellectual autonomy" (Cobb, 1988, p. 99). The TDS, by contrast, is oriented towards the diffusion of mathematical knowledge. The TDS is embedded in a social project whose goal is to make the students acquire a constituted knowledge (Brousseau, 2003). As a result, in the TDS, the emphasis has generally been put on mathematical knowledge and the efficient management of the learning environment. In constructivism, the emphasis has generally been put on the knowing subject—more specifically, on the understanding of the idiosyncratic manners in which students build their own knowledge. In the first case, the underpinning theoretical orientation has been essentially epistemological. In the second case, the theoretical orientation has been psychological. *The theory of objectification is located in a different educational project: it posits the goal of Mathematics Education as a political, societal, historical, and cultural endeavour aimed at the dialectical creation of reflexive and ethical subjects who critically position themselves in historically and culturally constituted mathematical discourses and practices, and who ponder new possibilities of action and thinking.*

The theory rests on the fundamental idea that learning is both about knowing and becoming. Behind this fundamental idea is the neo-Hegelian, dialectical, dynamic, constitutive conception of subjects and cultures: both the individual and culture are coterminous entities in perpetual flux, one continuously becoming the other and the other the one.

Let us turn now to the second chapter of this book where I identify the place of the theory of objectification in the landscape of contemporary sociocultural mathematics education theories.

Notes

1 As Walkerdine noted, "In the work of Piaget an evolutionary model was used in which scientific and mathematical reasoning were understood as the pinnacle of an evolutionary process of adaptation. The model viewed the physical world as governed by logicomathematical laws, which came to form the basis of children's development of rationality" (Walkerdine, 1997, p. 59).

2 The continuous dialogue of the theory of objectification with constructivism in general, with the TDS (Brousseau, 2002), with the commognition (Sfard, 2008), and, more recently, with the Espace de Travail Mathématique (Kuzniak, Tanguay, & Elia, 2016a; Radford, 2016a) and Inferentialism (Derry, 2017; Radford, 2017b) is not intended to create polarities. It is an effort to better understand our own theory. Vygotsky suggested that "Through others, we become ourselves" (1998, p. 170). The same is true of theories. It is through dialogue with other theories that theories form an identity.

CHAPTER 2

An Overview of the Theory of Objectification

1 Outline

In this chapter I present an overview of the theory of objectification. I start with a succinct historical account of the emergence of sociocultural research in mathematics education. This historical account paves the way to better understand the place of the theory of objectification within the landscape of current sociocultural research. The historical account also offers a background for a discussion of some of the basic conceptual orientations of the theory including a reconceptualisation of key educational concepts such as learning, and teachers and students. To better grasp the general conceptual orientation of the theory, I also discuss in this chapter the role it ascribes to artefacts, signs, language, and embodiment. The ideas introduced here will be addressed more comprehensively in the rest of the book, particularly in the more "theoretical" chapters, when I introduce in detail other key concepts (e.g., the concept of knowledge and knowing, and the concept of being and becoming).

2 Introduction

The theory of objectification is part of the increasing field of contemporary educational sociocultural theories (e.g., Atweh, Forgasz, & Nebres, 2001; Bartolini Bussi & Mariotti, 2008; Cantoral, 2013; Lave, 1988; Lave & Wenger, 1991; Sfard, 2008; Saxe, 2012; Stetsenko, 2017) whose common denominator is the claim that human beings are *consubstantial* with the culture or cultures in which they live their lives. What consubstantiality means is that there is a profound intertwining of cultures and what their individuals think, do, feel, imagine, hope and dream.

Contemporary sociocultural theories are the result of a historical millennial process of making sense of the Other. This process became more salient during the late Middle Ages in the commercial exchanges with China and Mongolia, then with the 15th century European colonising expeditions. It is in the course of these encounters with other cultures that conquerors, merchants, officials, and priests found themselves immersed in different worlds populated with individuals having different worldviews and values. The encounter with the Other was first (and, to a great extent, continues today to be) an experience

 | DOI: 10.1163/9789004459663_003

of assimilation: the assimilation of "they" to "us." In his book, *The conquest of America: The question of the Other*, one of the most famous scholars of Bakhtin's (1981, 1986) work, the Bulgarian-French literary critic Tzvetan Todorov, notes how Christopher Columbus systematically assimilates the aboriginal words to the semantics of the European worldview. Todorov writes: "the entire dimension of intersubjectivity ... of the human and therefore arbitrary character of signs, escapes him" (1984, p. 29). Todorov reports a revealing episode in which

> having learned the Indian word *cacique*, he [Columbus] is less concerned to know what it signifies in the Indians' conventional and relative hierarchy than to see to just which Spanish word it corresponds, as if it followed of itself that the Indians establish the same distinctions as the Spaniards, as if the Spanish usage were not one convention among others, but rather the natural state of things. (Todorov, 1984, p. 29)

The 20th century was witness to a series of efforts to understand the Other in new terms (e.g., Said, 1994) and the role of culture in the constitution of the mind and the self (see, e.g., Lévi-Strauss, 1962; Lévy-Bruhl, 1922, 1928; Luria 1931, 1934, 1979; Tulviste, 1991). One of the greatest obstacles in this endeavour has been to overcome the assimilative view—epitomised by Columbus—that reduces the others, their languages, their values, and understandings of the world to a given worldview.

In the field of mathematics education, sociocultural theories emerged during the 1980s and 1990s. The driving force was the pressing need to offer alternatives to:

1. the individualist learning approaches that have dominated the field (see Lerman, 1996a, 1996b), and
2. the Eurocentric conception of mathematics (see e.g., Bishop, 1988; D'Ambrosio, 1985).

To respond to these general needs, some mathematics education researchers tried to come up with socially, historically, and culturally oriented conceptions of the mind and a broader conception of psychology than the narrow, idealist, subject-centred one inspired by the natural sciences. These mathematics education researchers tackled, in different ways and from different perspectives, the question of the consubstantiality of the individuals and their culture, and ended up focusing on different problems—e.g., language and discourse, material culture and its effect on the individual's cognition, the sociocultural and political context of mathematics teaching and learning, and the ensuing manners of doing mathematics and thinking mathematically.

This range of problem is, indeed, what transpires in an overview of socio-cultural research published in the meetings of the International Group for the Psychology of Mathematics Education (PME) from 1978 to 2005. In this overview, Lerman identified four crucial categories:

1. Cultural psychology, including work based on Vygotsky, activity theory, situated cognition, communities of practice, social interactions, and semiotic mediation;
2. Ethnomathematics;
3. Sociology, sociology of education, poststructuralism, hermeneutics, and critical theory; and
4. Discourse, to include psychoanalytic perspectives, social linguistics, and semiotics (Lerman, 2006, p. 351).

Although each one of these categories cannot be considered as strictly separate or independent from the others, we can recognise early representatives of the first category in the works of Bartolini Bussi (1991), Bishop (1985), and Lerman (1992). Two representatives of the PME research in the ethnomathematics category are Bishop and Pompeu (1991), and Lea (1990), while Atweh and Cooper (1991), on the one hand, and Ellerton (1991), on the other hand, are representatives of Lerman's third and fourth categories, respectively.

In different ways, these researchers were interested in understanding the role of culture, history, and society in conceptions of mathematics and in ideas about the students' learning—conceptions and ideas that we are still trying to understand and that are far from being settled. Bartolini Bussi (1991) and Bartolini Bussi and Mariotti (1999, 2008), for instance, resorted to Vygotsky's work to investigate the role of the interaction and the role of cultural artefacts in the classroom; they called attention to the problem of semiotic mediation. Arzarello and collaborators focused on artefacts too, and more specifically, on mathematical signs and their evolution (Arzarello, Bazzini, & Chiappini, 1994). Inspired also by the work of Vygotsky (1987) and discursive psychology (Harré & Gillet, 1994), Lerman (1996b) studied the role of language in the constitution of intersubjectivity. Boero, Pedemonte, and Robotti (1997) investigated the students' emerging mathematics conceptualisation as a phenomenon at the crossroad of phylogenesis and ontogenesis through the historicity of speech.

These pioneering studies brought a different conception of language, interaction, and material culture. The work of Boero and his team, for example, turned to the work of Bakhtin (1986) and his concepts of language and intersubjectivity. Bakhtin's concepts of language and intersubjectivity are certainly at odds with traditional psychology's concept of language and its idea of the individual as an autonomous entity who comes into the world somehow endowed with the internal cognitive mechanism necessary for its intellectual development (Martin, 2004). Morgan (2006, 2018) resorted to social semiotics, while Sfard (2001), drawing on Vygotsky, suggested that thinking—one of those human features traditionally considered to be the most private and personal—has its roots in the social plane of communication: "thinking," Sfard (p. 26) says, "arises as a modified private version of interpersonal communication."

Where does the theory of objectification stand vis-à-vis those sociocultural trends? To answer this question, in the next sections I sketch the theoretical underpinnings of the theory of objectification.

3 Theoretical Underpinnings of the Theory of Objectification

I begin this section with a discussion of the concepts of teachers and students as conveyed by the theory of objectification. Then, I move to language-centred types of theorising in mathematics teaching and learning. This is followed by a discussion of the concept of activity as articulated in dialectical materialism. Such a concept appears as the central concept in the theory of objectification. The concept, however, is reformulated as *joint labour* for reasons that are explained below.

3.1 *Teachers and Students*

In the previous chapter I referred to a Grade 4 class episode and asked the question: If we are interested in providing accounts of the students' learning, to what exactly should we attend? Within a Piagetian pedagogical paradigm, the answer would be: the students. More precisely, it would be the students' *actions and utterances*, for in such pedagogical paradigm learning is precisely that which emanates from what the students *do and say*. The theory of objectification takes a different route. It attends to the students, but in a manner that is different from the individualist accounts of the child-centred pedagogy.

The problem of moving to a different conception of the student is certainly not easy: we come from a long historical tradition that considers student learning as the outcome of the student's own deeds. We find this idea clearly articulated in the "Universal Teaching" pedagogic program of the French educator Joseph Jacotot (1770–1840) in the 19th century. Jacotot's fundamental idea was that any individual can learn anything by herself. We do not need to teach the students. They will learn by themselves. As Jacotot (1828, p. 25) tells his students, "You do not know integral calculus, but you can learn it without explanations." This conception of learning as the outcome of the student's own deeds, which, as we saw in the previous chapter, was endorsed by constructivist pedagogies, has become a kind of second nature in mathematics education.

This is why the teacher's intervention in line 29 of the Piggy Bank example (Chapter 1) may feel disturbing. Indeed, as noted at the end of Chapter 1, the teacher's intervention may seem intrusive and as interfering with the students' conceptualisations and learning.

What if we put on hold for a second the idea that concepts, meaning, and intentionality necessarily have to emanate from the student in order for her to genuinely understand and learn mathematics? I am not advocating—and I should rush to clarify it—a pedagogy in which the teacher tells the students how to solve the problems and think mathematically. What I am saying is that things may be much more complex. Unfortunately, education research and practice have narrowed the teacher/students relationship to two possibilities:

(A) either learning emanates from the students or
(B) the students receive knowledge from the teacher.

This "either (A) or (B)" position is what Lerman (1996b) calls in his critique of constructivism the *absolutist alternative*. Following Lerman, I claim that the absolutist position leads to considering the teacher/students relationship in a too simplistic manner.

Possibility (B)—that students receive knowledge—supposes that knowledge can be transmitted from one individual to another (from a sender to a receiver). This is the main idea of the transmissive pedagogical model—and is indeed a very naive one. It resorts to a very crude idea of knowledge: knowledge is conceptualised as something that individuals possess or not, hence to a reified or commodified concept of knowledge. It also resorts to a very superficial and problematic idea of the teacher and the student.

Within the absolutist position, the only possible solution seems then to be possibility (A): that knowledge has to *start* and *result* from the students' deeds. This is the position that Williams (2016) adopts in his critique of Roth and Radford's (2011) book. Working within the borders of the absolutist position, Williams opts for possibility (A). The conclusion is not hard to guess: he ends up adopting the ideas of knowledge and the student of constructivism and Piagetian pedagogy (see Radford & Roth, 2017, for a detailed analysis).

Williams's critique is perhaps a token of the difficulties that result from trying to overcome the absolutist position. Already in his early work, Vygotsky claimed a crucial role for the student in her own learning, without, however, reducing learning to the students' deeds. Thus, the old pedagogy (i.e., the transmissive pedagogy), Vygotsky said, "treated the student like a sponge which absorbs new knowledge." Elaborating on the idea in more detail, he claimed that "the assumption that the student is simply passive ... is the greatest of sins, since it takes as its foundation the false rule that the teacher is everything and the student nothing" (Vygotsky, 1997b, p. 48).

From a Vygotskian perspective, the course of action lies not in choosing between (A) *or* (B), nor is it about keeping both, as Godino and Burgos (2020) suggest in their article, *Between Transmission and Inquiry*. Indeed, instead of (A) *or* (B), Godino and Burgos suggest (A) *and* (B). Because of the complexity of mathematical knowledge, the student's autonomy, they argue, cannot be the starting point. We can start from (B), with a "minimal teaching guide," (p. 96) and then move to (A) through a process that ends with the students' acquisition of institutional knowledge. Following Vygotsky's ideas, the theory of objectification is an attempt to move beyond the absolutist position and each one of its terms and combinatorial possibilities.

In this line of thought, let us agree to put on hold the idea that learning has to come from the students *or* from the teacher. Then, the classroom episode referred to in the previous chapter appears in a new light. It appears much more *complex*. It requires us to see the teacher's intervention and the students' reactions through new lenses. The interaction of the teacher and the students appears as a dance of consciousness trying to grasp something *together*.

Naturally, this new look at the teacher-and-students' interaction demands to see teaching and learning in a different way. Not only does the concept of the student have to be revisited, but also the *concept of the teacher*. The teacher, Vygotsky argued, cannot be pictured as injecting knowledge into the student's mind. Indeed,

> Just as a gardener would be acting foolishly if he [sic] were to try to affect the growth of a plant by directly tugging at its roots with his hands from underneath the plant, so is the teacher in contradiction with the essential nature of education if he [sic] bends all his efforts at directly influencing the student. (Vygotsky, 1997b, p. 49)

A few years ago, Brent Davis, in a conference in Toronto, pointed out that the focus of mathematics education has shifted historically: from problems about curriculum organisation we went to problems concerning the child's mathematical thought; and from there we moved to problems concerning the teacher. This is where we are at today. Yet, to a large extent, the teacher remains considered along the lines of a technical agent—a kind of industrial operator—whose mission is to deliver a curriculum (Brown, 2010; Grootenboer & Jorgensen (Zevenbergen), 2009; Pais, 2013; Valero, 2004). This is a very restrictive view of the teacher, one that Canadian educator Ted Aoki criticised by reducing the teacher to "a competent teacher-implementer"

> who has skills and techniques oriented toward efficient control. Such a know-how-to-do view of implementation is embedded in scientific and

> technological thought/action framework that reduces human competence to instrumental reason and instrumental action. Here, the teacher is seen as a rule-oriented, rule-governed being cast within a manipulative ethos, an ethos in which even his [or her] future is conceived in terms of rules. (Aoki, as cited in Pinar & Irwin, 2005, p. 113)

The concept of the teacher and the student that the theory of objectification brings with it draws on an anthropological perspective that I present briefly below—one in which teachers and students are not portrayed as self-sufficient and self-made beings who essentially already know their business. More specifically, the student is not considered as a rational cogitator or epistemic subject, nor is she reduced to a cognitive subject. Likewise, the teacher is not considered a technocrat who is there to help or to coach the students, someone who dispenses knowledge through scaffolding or other strategies. Nor is the teacher considered a *mediator*, as often suggested by sociocultural theories. One of the problems with this patriarchal view of the teacher (helper, coach, assistant, mediator, etc.) is that the teacher is imagined as an already given entity of the teaching and learning process. That is, the teacher appears as a benevolent detached participant. This is not what I see either in our pre- and in-service teacher education programs or in the classrooms where I conduct my research. What I see are teachers who struggle along with their students in order to get things done. Thus, instead of being considered as entities already given, as sources of knowledge and intentionality, students *as well as teachers* are considered as subjectivities in the making, openness towards the world. Teachers and students are conceptualised as human beings in flux, as unfinished and continuously evolving projects of life, in search of themselves, engaged together in a *same* endeavour where they suffer, struggle, and find enjoyment and fulfillment together.

Seeing the Piggy Bank example through the lenses of the theory of objectification, Mrs. Giroux—the teacher—appears engaged with Albert, Krysta, and Manuel. Mrs. Giroux is not doing the same thing as the students. Yet, they are *labouring together*, trying to bring to the fore a form of co-variational thinking about sequences. What they are doing is the creation of what Hegel (2001) called a *common work*.

In the theory of objectification common work is defined as the sensuous appearance of knowledge—that is, the sensuous appearance of a co-variational algebraic way of thinking through collective problem posing and solving, and discussion and debate in the classroom. The teacher's and students' *common work* is the bearer of dialectical tensions because of the emotional and conceptual contradictions of which it is made. Through it, knowledge appears sensuously in the classroom (through action, perception, symbols, artefacts, gestures, language), much in the same way, and with similar aesthetic force, that music appears aurally in a concert hall through the common work of the members of the orchestra (Radford, 2019).[1]

This view of the teacher and the students is in harmony with the general Freirean educational project in which the theory of objectification is based and that I mentioned at the end of the previous chapter: a political, societal, historical, and cultural endeavour aiming at the dialectical creation of reflexive and ethical teachers and students who critically position themselves in historically and culturally constituted mathematical discourses and practices, and ponder and deliberate on new possibilities of action and thinking. What the theory of objectification attends to is hence not only the realm of *knowledge* but also the realm of *becoming*. That is, in the theory of objectification the focus shifts from how students receive knowledge (transmissive teaching) and how students construct their own knowledge (constructivism), to how teachers and students, working together, produce knowledge in the classroom against the backdrop of culture and history. But the focus also shifts to how teachers and students co-produce themselves as subjects, in general, and as subjects of education, in particular.

3.2 *Language*

I have just said that one of the specificities of the theory of objectification is its focus on the teachers *and* the students in providing accounts of teaching and learning. What about language? Language is certainly an important element in the theory of objectification. Yet, its theoretical positioning is different from what we find in other sociocultural research. Let me explain. I mentioned above that language was one of the theoretical orientations at the base of the emergence of sociocultural theories in mathematics education. The interest in language has not diminished. On the contrary, in the past two decades, language has become one of the most prominent research areas in mathematics education (Radford & Barwell, 2016). And such prominence should not be a surprise.

Indeed, at the heart of sociocultural theories developed in anthropology and sociology we find a constant struggle to understand the problem of the relationship between the individual and the social, the subjective and the objective, and inner and outer life. It is in this context that, in the work of Wilhelm von Humboldt (1998), Ernst Cassirer (1980), Mikhail Bakhtin (1986), and Valentin Voloshinov (1973), language came to be considered the link between cultures and their individuals. Bakhtin (1986), for example, turned to literature and speech genres. He saw in the historically developed speech genres the forms through which individuals relate to their culture. Speech genres offer to the individuals social locations from where to see, interpret, and speak. Voloshinov (1973, also spelled as Vološinov) turned to the sign. He considered the sign to be both *subjective* (in the sense that it is produced by an individual and expresses the individual's intentions) and *objective* (in the sense that the expressivity of the sign is embedded in a cultural system of expressions and values). The sign ensures an entanglement between the subjective and the objective so that "the very notion of a fundamental, qualitative difference between the inner and the outer element [of life] is invalid to begin with" (Voloshinov, 1973, p. 85). In Voloshinov's account the sign par excellence is the *word*. The "word is present in each and every act of understanding and in each and every act of interpretation" (Voloshinov, 1973, p. 15). According to Volshinov, "the word is implicated in literally each and every act or contact between people" (p. 19). Drawing on this conception of the word, Voloshinov, as well as Bakhtin (1986), came to see language and literature not just as one of the fields of aesthetic experience and cognition but as the *central* field through which the other cultural fields are refracted. For them, it is ultimately through language and literature that reality is produced and interpreted.[2] The problem of the individual and the social, the subjective and the objective, and the inner and outer life is resolved, in Voloshinov's account, in the dialectical tension between the relatively stable centripetal forms of culture (epitomised by the novel) and the centrifugal forms of resistance and novelty (epitomised in Bakhtin's idea of carnival). Yet, as Tihanov notes,

> Vološinov ended up reducing culture to a single-ideal-mode of existence ... [one in which] language had a conspicuous and autonomous role in the formation of social life. Thus ... Vološinov pursued a language-centered type of theorizing in the field of social thought. (Tihanov, 1998, p. 615)

In the theory of objectification language remains an extremely important component. Language appears as a powerful means in knowing and becoming. Yet, it is not the central organising ontological category as is the case with other sociocultural approaches (e.g., Harré & Gillett, 1994; Sfard, 2008). The ontological category of the theory of objectification is not language but *activity* (understood as *joint labour*).

The question of whether or not a Vygotskian or, more generally, a dialectical materialist theory based on language and discourse and one based on the concept of activity are equivalents is still a much discussed point among sociocultural theorists.[3] I do not think that both choices lead to equivalent theories. I do not think that they produce the same account of learning and teaching or the same account of intersubjectivity. For one thing, the unit of analysis is not the same, nor are the methodological approaches through which the unit of analysis is put under scrutiny. I do think, though, that both kinds of (dialectical materialist) theories bear a distinctive theoretical affinity. This affinity will be greater or lesser depending on how discourse and activity are respectively conceptualised. Before I turn to the precise definition of activity in the theory of objectification, let me say something about the difficult question of the role of signs and artefacts in the social formation of the mind.

3.3 *Signs and Artefacts*

Sociocultural theories have extensively argued for the distinctive role of signs and artefacts in human life and cognition. Wartofsky (1979, p. 202), for instance, claimed that "what constitutes a distinctively human form of action is the creation and use of artifacts, as tools, in the production of the means of existence and in the reproduction of the species." A shared idea is that signs and artefacts are bearers of meanings produced by the intellectual activity of previous human generations (Baudrillard, 1968; Borba & Villareal, 2006; Leontiev, 1968; Malafouris, 2004; Meira, 1998; Vygotsky, 1929). Human intelligence is embedded and embodied in them. Yet, the embodiment of human intelligence in material objects or in language/discourse is not enough for this intelligence to be automatically disclosed to the individuals. The measuring tape that we use in construction settings is a good example. The work of LaCroix (2014) shows how difficult it is for pre-apprentices enrolled in pipe trade training to make sense of the arithmetic of $(\frac{1}{2})^n$ behind the measuring tape. Once the pipe trade pre-apprentices become acquainted with such an arithmetic, a new world of possibilities for thought and action is disclosed.

Now, despite the widespread agreement with the distinctive role of signs and artefacts in human life and cognition, their precise role in human development has been a controversial question—even in the evolution of a *same* sociocultural theory.

Take the case of Vygotsky. We know, indeed, that in the late 1920s, trying to understand human cognition from a broader perspective than the one offered by the reflexology research and rational psychology of his time, Vygotsky stressed the impact of the use of signs on human thought. Signs for him—in contrast to other semiotic approaches, such as Peirce's (1931–1958)—were not considered as means to *represent* things. Influenced by his work on special education Vygotsky developed a non-representational conception of signs (Vygotsky, 1993). In an article featuring the question of the child's cultural development, Vygotsky noted that

> From a pedagogical point of view, a blind or deaf child may, in principle, be equated with a normal child, but the deaf or blind child achieves the goals of a normal child by different means and by a different path. (Vygotsky, 1993, p. 60)

How? Through education, by "creating artificial, cultural techniques, that is, a special system of cultural signs and symbols" (Vygotsky, 1993, p. 168). In other words, auxiliary material cultural means (e.g., Braille dots) compensate for differences in the child's sensorial organisation. Vygotsky thought of these compensating means as signs.

But it would be a mistake to consider signs as merely compensating devices. For Vygotsky, the most important point was that, through signs, we come *to control our psychological behaviour*. In other words, signs serve to fulfill psychological operations and in doing so they *alter* human cognition (Radford & Sabena, 2015). Thus, in a passage from *Tool and Symbol in Child Development*, Vygotsky and Luria argued that

> the use of signs, which are at the same time a means of auto-stimulation, leads man [sic] to a completely new and specific structure of behaviour, breaking away from the traditions of natural development and creating for the first time a new form of cultural psychological behaviour. (Vygotsky & Luria, as cited in van der Veer & Valsiner, 1994, p. 145)

It is in this context that Vygotsky formulated his concept of sign, his famous "genetic laws of cultural development," the concomitant and controversial idea of internalisation (to which I shall return in Chapter 4), and his concept of semiotic mediation. These concepts were the central theoretical categories in Vygotsky's study of the mind around the period that roughly extends from 1927 to 1931. These concepts belong to what contemporary Vygotskian scholars call Vygotsky's instrumental "phase" or "moment" (for details see Presmeg, Radford, Roth, & Kadnuz, 2016). Because of the analogy of the sign with the labour tool, González Rey (2009) qualifies the 1927–1931 moment as an *instrumentalist* turn.

However, at the end of his life, Vygotsky was changing his mind about the role of signs. He started moving towards a broader research program. While signs and tools remained present, they however lost their technological flavour of the instrumentalist phase, "where the psyche was still viewed by analogy with a machine, a technical device" (Zavershneva, 2010, p. 72). Signs and tools became considered in terms of what they *signify*. Vygotsky was moving from an instrumentalist conception of signs to an aesthetic one, yielded room to a programmatic study of consciousness in which affect, emotions, and meaning were brought to the fore. Vygotsky noted: "Consciousness as a whole has a semantic structure. We judge consciousness by its semantic structure ... Semiotic analysis is the only adequate method for the study of the systemic and semantic structure of consciousness" (Vygotsky, 1997c, p. 137).

In some manuscript notes from 1934, Vygotsky writes: "[T]here are 2 unities of dynam[ic] activity: think[ing] and real activity ... In fact, the role of think[ing] in activity consists of introducing new dynam[ic] possibilities to activity" (as cited in Zavershneva, 2010, p. 83). Within the program of research that Vygotsky was envisioning at the end of his life, thinking and activity became intertwined. Such a move does not diminish the importance of signs and artefacts. Signs and artefacts appear now subsumed into a different and more encompassing concept: the concept of *activity*. This is the conceptual stance that the theory of objectification takes on the role of signs and artefacts. Signs and artefacts do not appear as *mediators* of activity, as is the case in other sociocultural approaches: they are rather considered an *integral part* of human thinking and human activity.

3.4 *Activity*

In subjectivist accounts of knowing, knowledge is the outcome of the individual's activity. Knowledge (K) is seen, hence, as an *extension* of the subject (S). If we want to express the idea through an equation, we can write: K = S. To produce knowledge, the subjective activity of the subject is enough. By purely and simply identifying itself with knowledge and its objects, by positioning itself as the epistemological foundation, the subject presents itself as a totality, a whole (Ricard, 1999, p. 279). This is not the theoretical position assumed in the theory of objectification. In the theory of objectification, the starting point is that, in learning settings (e.g., a classroom), knowledge (cultural knowledge) is created and recreated through sensuous cultural-historical activity. Knowledge can come into sensuous existence only *in* and *through* practical, collective activity; that is, activity with others. For, as the pre-apprentices of the pipe trade know very well, the human intelligence embodied in the measuring tape does not show up in a direct and immediate fashion; nor can it be reconstructed on the basis of personal experience alone. Indeed, no personal experience, regardless of how rich it might be, can manage to establish on its own a cultural system of ideas—such as the arithmetic of $(\frac{1}{2})^n$ behind the measuring tape, the Roman legal system, formal logic, or set theory. An undertaking of this magnitude would require not one lifetime, but thousands (Leont'ev or Leontiev, 1968), or perhaps it cannot be reconstructed at all. In order for learning to occur, a specific human endeavour is hence required: *activity*. The classroom production of knowledge and its progressive disclosing to the students' consciousness is entangled with the evolving classroom activity that makes such a production/ disclosing possible. As a result, the *kind* of mathematical activity that we can foster in the classroom becomes extremely important.

Now, the activity to which I am referring here has a definite sense that is very different from the usual conceptions that reduce activity to a series of actions that an individual performs in the attainment of his or her goal. The latter line of thinking reduces activity to a functional and *technical* conception: activity amounts to the deeds and doings of the individuals. Activity in the theory of objectification does not merely mean to do something. Activity (Tätigkeit in German and deyatel'nost' in Russian) refers to a *dynamic system* where individuals interact collectively in a *strong* social sense, which makes the activity's products collective as well. This is why activity as Tätigkeit should not be confounded with activity as Aktivität/aktivnost'; that is, as simply being busy with something (Roth & Radford, 2011).[4] Activity as Tätigkeit is a joint endeavour through which individuals produce their means of subsistence. It "comprises notions of self-expression, rational development, and aesthetic enjoyment" (Donham, 1999, p. 55). More precisely, activity as Tätigkeit is a *form of life*. It is the endless process through which individuals inscribe themselves in society (Presmeg, Radford, Roth, & Kadnuz, 2016). To avoid confusion with other meanings, in the theory of objectification, activity as Tätigkeit is termed *joint labour*.

3.5 *Joint Labour*

Joint labour is the chief ontological category of the theory of objectification. I would like to insist on the fact that this theoretical choice does not dismiss the role of language, signs, and artefacts in the processes of knowing and becoming. This choice rather asserts the fundamental ontological and epistemological role of matter, body, movement, action, rhythm, passion, and sensation in what it is to be human.

Let me attempt to clarify further the choice of activity as the main organising category and its relation to language. In a short note, the American cellist Yo-Yo Ma writes about the 2018 release of his *Six Evolutions—Bach: Cello Suites*. He says: "For almost six decades, they [Bach's cello suites] have given me sustenance, comfort, and joy during times of stress, celebration, and loss ... I share this music, which has helped shape the evolution of my life."[5] Language makes it possible for Ma to give conceptual content to a series of emotions, such as comfort and joy, celebration and loss, that otherwise would remain merely sensed. Through language he can *recognise* them and *name* them. Through language he can share these emotions. But these emotions are tied to things, events, and individuals that were part of various significant *activities* in Ma's life—activities that certainly *included* language and discourse but were not reduced to them.

In the early 1940s, Leont'ev was developing Vygotsky's late ideas about the centrality of activity in understanding human thinking. In a lecture from 1940, published later with the title, *The Genesis of Activity*, Leont'ev emphasises the importance of language and communication, and goes on to say that "Communication is the aspect of labor that makes speech essential, but speech itself emerges in direct connection with labor itself, and therefore speech itself becomes possible only in the process of labor" (Leontiev, 2005, p. 69).

Activity (or joint labour with other people, in our terminology) *includes* language but also includes in a decisive manner the embodied experiences of movement, action, rhythm, passion, and sensation.

Coming back to our educational discussion, at a practical level, the concept of joint labour allows one to revisit the concept of classroom teaching and learning activity and the role of language, signs, and artefacts therein.

On the one hand, as mentioned previously, language, signs, and artefacts are not considered as mediators of activity. They are rather considered as part of the individuals' activity. And since thinking and activity are considered here as intertwined (Vygotsky, as cited in Zavershneva, 2010), language, signs, and artefacts are also considered as part of thinking. They are part of the *material texture* of the individuals' thinking.

On the other hand, the concept of joint labour allows one to conceive of classroom teaching and learning not as two separate activities, one carried out by the teacher (the teacher's activity) and another one carried out by the student (the student's activity), but as a single and same activity: the same teachers-and-students joint labour. If we come back to the Piggy Bank example discussed in the previous chapter and see it through the lenses of the theory of objectification, the teacher does not necessarily appear as a possessor of knowledge who is delivering or transmitting knowledge to the students nor as someone scaffolding strategies to the students. Nor do the students appear as passive subjects receiving knowledge. In the Piggy Bank example, the teacher and the students appear as *labouring together* towards the production of a *common work*: in the Piggy Bank example, the common work is the sensuous appearance of a co-variational way of thinking about sequences. It is in the production of this common work that the students are conceived of as encountering and becoming gradually aware of culturally and historically constituted forms of mathematics thinking.

3.6 *Processes of Objectification*

I mentioned above that subjectivist accounts of knowing can be summarised through the equation K = S. The subject, S, produces her own knowledge, K.

This is not the case in the theory of objectification. Before our encounter with knowledge, knowledge presents to us as something *different* from us: something that in its *alterity*, in its otherness, its own presence *objects* us; that is, resists or opposes us. The equation is: S ≠ K. Before our encounter with knowledge, knowledge is the sign of a *difference*. *Object*-ification is the attempt to erase that difference. It happens in joint labour. The joint labour-bounded *encounters* with historical forms of mathematics thinking are termed *processes of objectification*. They are progressive processes of encounter that are sensible and sensuous, ideational and material, symbolic and discursive, affective and emotional, critical and subversive, and transformational and poetic.

In the next chapters I will present an operational definition of objectification along with some concrete examples. In the next subsection I dwell on the anthropological conception of the human.

3.7 *The Anthropological Conception of the Human*

Humans, following Marx's (1988) Spinozist stance (see Spinoza, 1989), are considered to be *part of nature*: they are natural beings. That humans are natural beings means that they are sensible beings, unavoidably affected by other parts of nature. In this context, sensations and passions are conceptualised as ontological affirmations of the individual's nature as natural beings (Fischbach, 2014).

One important consequence of this theoretical stance is that individuals cannot be conceived of as substantial entities, produced from within, as articulated by the humanist trend of the Enlightenment that informs constructivism and other contemporary educational theories. Individuals are *relational* entities through and through. They are profoundly linked to an ensemble of relationships with other parts of nature—including social relationships—and their existence is based on culturally and historically constituted conditions of life. In this line of thought, to be a natural being means also that, like other natural living beings, humans are *beings of need* who find their satisfaction in objects *outside* of themselves.

To meet their needs (needs of survival as well as artistic, spiritual, and other needs created by/in society), humans engage themselves actively in the world. They *produce*. What they produce to fulfil their needs occurs in a social process that is, at the same time, the process of the individuals' inscription in the social world and the production of their own existence. The name of this process is what I have termed joint labour. Sensuous, material joint labour is considered the ultimate field of aesthetic experience, subjectivity, and cognition.

The previous ideas bring me to the question of embodiment.

3.8 *Embodiment*

As in the case of other central concepts (such as activity and material culture discussed above), embodiment in the theory of objectification takes a precise meaning. It does not refer to the material body that is discussed in other approaches to mathematics teaching and learning (for a critical overview of embodiment in mathematics education, see Radford, Arzarello, Edwards, & Sabena, 2017). Embodiment is related to the idea of *sensuous cognition* (Radford, 2013a, 2014a), which rests on a specific historical understanding of sense, sensation, materiality, and their entanglement with the conceptual realm. Within this theoretical perspective, our cognitive domain can only be understood as a culturally and historically constituted sentient form of creatively responding, acting, feeling, imagining, transforming, and making sense of the world.

The historical nature of the senses and sensation and the entanglement of cognition and the world make it impossible to reduce the material world to pure stuff. Matter, notwithstanding the empiricists, is not merely the concrete stuff that we touch with sensuous certainty with our hands or perceive with our eyes, something as banal as, for example, a cherry tree. Matter—all that is revealed to us in our dealing with the world—is cultural-historical in the sense that it bears in itself the traces of human labour and intellectual activity. To see matter only subjectively is to make the mistake for which Marx reproached materialist philosopher Ludwig Feuerbach (1966):

> [Feuerbach] does not see that the sensuous world around him is not a thing given direct from all eternity, remaining ever the same, but the product of industry and of the state of society; and, indeed, [a product] in the sense that it is an historical product, the result of the activity of a whole succession of generations ... Even the objects of the simplest "sensuous certainty" are only given him through social development, industry and commercial intercourse. (Marx, 1998, p. 45)

What Feuerbach does not grasp, Marx contends, is that what we access through our senses is not the world in its intact materiality, Adam and Eve's intact world, but a world of labour replete with historical and cultural sensuous activity:

> The cherry-tree, like almost all fruit-trees, was, as is well known, only a few centuries ago transplanted by *commerce* into our zone, and therefore only by this action of a definite society in a definite age it has become "sensuous certainty" for Feuerbach. (Marx, 1998, p. 45, original emphasis)

The concept of sensuous cognition rests on this understanding of sense, sensation, and the material world as historical and cultural products.

However, the concept of sensuous cognition that provides embodiment with a specific meaning in the theory of objectification goes beyond asserting that mind, body, and world are dialectically intertwined entities. It also asserts that *cognition and the human senses are transformed by human labour and social practice*. That is, human cognition and our body and senses are not a natural phenomenon, but a cultural-historical one.

As we can see in the Piggy Bank example in the previous chapter, the teacher and the students have recourse to gestures and the material culture in a decisive manner. Gestures, the goblets, the bingo chips, their colour, etc., are crucial in the unfolding of joint labour. Through them the teacher and the students infuse the bit-by-bit appearing co-variational manner of perceiving, intuiting, and thinking about sequences with an emerging meaning. Yet, this emerging meaning can only be understood against the background of the teacher's and students' joint labour.

As in the case of signs and tools, in the theory of objectification, embodied actions (gestures, body posture, corporeal movement, perception, tactility, prosody) are not considered mediating devices of activity but parts of activity and part of the teacher's and the students' unfolding thinking.

3.9 *Variations on the Piggy Bank Example*

Let us pause for a moment and imagine the same Piggy Bank problem discussed in Chapter 1 in a different scenario. Let us imagine that the episode now unfolds according to the transmissive educational model of direct teaching. The teacher is in front of the class and tells the students how to solve the problem. The students listen and take some notes. Can we say that there is joint labour here? Let us repeat the same thought experiment, but this time the episode unfolds according to the constructivist student-centred model. The students are trying by themselves to solve the Piggy Bank problem and the teacher refrains from intervening. Can we say that there is joint labour here? In both cases, what teachers and students are doing is not something really *collective* in a *strong* social sense, geared to the satisfaction of a common need. Teachers and students are involved in classroom activity, but it is an activity in the sense of Aktivität/aktivnost'. More specifically, in the transmissive model, the teacher and the students are involved in a functional and technical coordination of actions: on the one hand, the teacher tells the students what to do; on the other hand, the students listen and try to make sense by themselves of what the teacher does and says. In the constructivist model, "the teacher and students each construct individual interpretations that they take as being shared with the others" (Cobb, Yackel, & Wood, 1992, p. 17). What results from such an activity cannot be a "common work" but something that is merely "taken-as-shared."

The Piggy Bank example presented in Chapter 1 provides us with an example of a process of objectification. Working together, the teacher and the students are producing (through gestures, posture, perceptual activity, language, artefacts) a *common work*, which allows the students to become progressively aware of a different manner to think about the problem (a co-variational way of thinking).

Through these social, material, embodied, and semiotic processes, the students and teachers not only create and re-create knowledge but they also co-produce themselves as subjects in general and as subjects of education, in particular. More precisely, they produce subjectivities; that is to say, singular individuals in the making. This is why, from this perspective, processes of objectification are at the same time processes of *subjectification*.

4 Summing up and Looking Ahead

To sum up this chapter, the theory of objectification is an attempt to understand teaching and learning not as the result of the individuals' own deeds (as in individualist accounts of learning) but as cultural-historical situated processes of knowing and becoming. It seeks to study the ways by which the students become progressively aware of historically and culturally constituted forms of thinking and acting, and how, as subjectivities in the making, teachers and students position themselves in mathematical practices. Such an investigation includes language, signs, artefacts, and the body, but the central theoretical construct is *joint labour*. Through the historical intelligence that they bear, signs and artefacts provide the individuals' joint labour with *organising possibilities* of action and thought. As part of the material culture, signs and artefacts are entangled with the body and cognition at large. But the body and cognition are understood here in their dialectical materialist sense: not as entities that evolve naturally; they are understood as culturally and historically constituted, embodied, and materially sentient forms of creatively acting, feeling, imagining, transforming, responding to, and making sense of, the world.

It is in this context that the concept of joint labour featured here acquires a specific meaning. In particular, as I discuss it in the next chapters, it resorts to *collective forms of knowledge production* and non-alienating *forms of human collaboration*. In contrast to other theories that focus only on learning mathematical contents and that see in classroom interaction a technical tool to precipitate learning, the theory of objectification seeks to provide the conditions not only for deep conceptual mathematical learning but also for the emergence of culturally evolved forms of human collaboration that may be conducive to the formation of critical subjectivities. As we shall see in the last chapter of this book, these forms of human collaboration rest on a communitarian ethics that includes, among other things, an *ethical commitment*. This ethical commitment is what we see in the Piggy Bank example when, for instance, in line 29 the teacher *engages* with the students and *invites* them to see the problem-solving strategy under a new light. The students *respond* to the teacher's call and, labouring with the teacher, try hard to make sense of what this new strategy might look like. I will come back to these ideas and try to elaborate on them in the rest of the book. In the meantime, Table 1 offers a synoptic panorama of what has been said so far.

TABLE 1 A synoptic panorama of the theory of objectification

	The *theory of objectification* (TO) is a teaching and learning theory inspired by dialectical materialism, Freire's concept of education, and Vygotsky's school of thought. Moving away from subjectivist (e.g., empiricist and constructivist) accounts of learning, and traditional subject-object epistemologies, it conceives of teaching and learning as a single process involving both *Knowing* and *Becoming*.
Keywords	– *Learning:* The TO conceptualises learning in terms of processes that are at the same time processes of objectification and processes of subjectification. – *Processes of objectification:* are the active, embodied, discursive, symbolic, and material processes through which the students *encounter*, *notice*, and become *critically acquainted* with culturally and historically constituted systems of thinking, reflection, and action. In this encounter, the students are faced with the alien, the *Other*. This encounter is sensed as the encounter of something that *objects* (etymologically speaking, something that is *set against* or that *opposes*) the individual. – *Processes of subjectification:* are based on the idea that we, humans, are always unfinished projects of life, subjects perpetually in the making. Processes of subjectification are the processes of the incessant making of the subject, of the continuous creation of a singular (and unique) historical and cultural subject. – *Mathematics*: is simultaneously ideal and concrete. Mathematics is visual, tactile, aural, material, artefactual, gestural, and kinesthetic—something produced by the joint labour of the teachers and the students.

(cont.)

TABLE 1 A synoptic panorama of the theory of objectification (*cont.*)

	– *Joint Labour:* is not a mere set of coordinated actions. It is the joint *activity* (*deyatel'nost'* in Russian) carried out by the teacher *and* the students, a form of *energy* embedding the flux of emotional, affective, ethical, intellectual, and material interrelated components out of which mathematics appears and objectification and subjectification occur. – Common Work: is defined as the sensuous appearance of knowledge—e.g., the sensuous appearance of a co-variational algebraic way of thinking through collective problem posing and solving, and discussion and debate in the classroom.
Basic tenets	– Educational settings do not merely produce knowledge but subjectivities too. – As a result, education should involve both *Knowing* (the dimension of knowledge) and *Becoming* (the dimension of the subject). – The TO considers Mathematics Education as a political, societal, historical, and cultural endeavour. Such an endeavour aims at the dialectic creation of ethical subjects who critically position themselves in historically and culturally constituted mathematical practices and ponder and deliberate on new possibilities of action and thinking.
Contributions	– Through fine-grained epistemological analyses the TO provides elements to create rich educational settings that promote the development of sophisticated classroom forms of knowledge production. Through the principles of a communitarian ethics that revolves around the ideas of solidarity, answerability, and the practice of empathy and the care of the other, the TO promotes non-alienating socioculturally oriented forms of human cooperation. In such settings, young students start progressively inscribing themselves in the social world while making sense of historically constituted complex systems of mathematical thinking (e.g., arithmetic, algebraic). Through lengthy processes of objectification and subjectification, the students engage in more and more elaborate forms of joint labour and increasingly complex structures of intersubjectivity (I-You, We-You, We-Them). Their multi-modal semiotic activity becomes re-organised and refined, giving rise to complex cultural forms of perceiving, imagining, languaging, symbolising, and thinking.

Acknowledgement

This is a revised and expanded version of Radford, L. (2018). A teoria da objetivação e seu lugar na pesquisa sociocultural em educação matemática. In V. Moretti & W. Cedro (Eds.), *Educaçao Matemática e a teoria histórico-cultural* (pp. 229–262). Campinas, São Paulo: Mercado de Letras. Revised with the permission of the copyright holder.

Notes

1 I come back to this idea in Chapter 3.

2 Bakhtin considered the novel as the most accomplished speech genre. Following the dialectical materialist view of reality as something continuously unfolding and moving, he argued that "The novel is the only developing genre and therefore it reflects more deeply, more essentially, more sensitively and rapidly, reality itself in the process of its unfolding. Only that which is itself developing can comprehend development as a process. The novel … best of all reflects the tendencies of a new world still in the making" (1981, p. 7).

3 For a discussion about the differences between discursive and other sociocultural approaches, see Stetsenko and Arievitch (1997). See also Markus's (1986) inspiring book, *Language and Production. A Critique of Paradigms*.

4 Engeström (1999, p. 21) complains that "the English term activity is unable to carry the deeper philosophical meaning of the original German concept of Tätigkeit."

5 See https://www.yo-yoma.com/sounds-symphonies-archive/six-evolutions-bach-cello-suites/

CHAPTER 3

Knowledge and Knowing

1 Outline

The previous chapters have laid the groundwork to address two key concepts of the theory of objectification: the concepts of knowledge and knowing. These concepts are necessary in order to define, in the next chapter, the concept of learning.

At first glance a discussion of knowledge and knowing might seem esoteric and futile. It seems to me, however, that if theories in mathematics education entertain the hope of offering coherent and cogent accounts of learning, they must first clarify what they understand by knowledge. Indeed, learning is always learning something: a content (e.g., in mathematics, learning can be about algebraic knowledge, geometric knowledge, etc.). As a result, accounts of learning cannot avoid addressing the question of knowledge and its nature.

In the following section I present a discussion on knowledge as it has generally been understood in mathematics education; that is, as *construction*. Then, I present the definitions of knowledge and knowing as they are understood in the theory of objectification. These sections are unavoidably theoretical. The chapter ends with a less theoretical section in which I make an analogy between mathematics and music, one that—I hope—can best help explain the meaning of the concepts of this chapter.

2 Knowledge

2.1 *Knowledge as Construction*

Generally speaking, mathematics education research does not really make a distinction between knowledge and knowing.[1] In this context, it is very common to speak of knowledge/knowing as something that one *makes* or *constructs*. Knowledge or knowing is to construct something.

The fundamental metaphor behind this idea is that knowledge/knowing is something similar to concrete objects in the world. One constructs or assembles knowledge as one constructs or assembles parts of a chair. This idea of knowledge/knowing as construction is relatively recent. It emerged gradually during the 16th and 17th centuries when manufacturing and the commercial

 | DOI: 10.1163/9789004459663_004

production of objects became the main form of human economic production in Europe.

Hanna Arendt summarises this modern idea of knowledge/knowing as follows: "I 'know' a thing whenever I understand how it has come into being" (Arendt, 1958, p. 585). It is thus within the general 16th and 17th centuries' view of a manufactured world that knowledge is first conceived of as a kind of constructed product. Arendt quotes the 18th century philosopher Giambattista Vico saying that "Mathematical matters we can prove because we ourselves make them" (p. 584). We find this same idea in Immanuel Kant's *Critique of Pure Reason*—a monumental 18th century book whose influence has not ceased to affect us. Kant tells us that "mathematics [...] derives its knowledge not from concepts but from the construction of them" (Kant, 2003, p. 590 (A 734/B762)).

This concept of knowledge as construction was featured by Piaget in his genetic epistemology and was widely adopted in mathematics education where an emphasis was put on the *personal* dimension of knowledge construction: you and only you can construct your own knowledge. In this view, knowledge is not something that someone can construct and pass on to another; what you know is the result of *your own* experience.

As many researchers have noted, such a view of knowledge is problematic for many reasons. For example, it leaves little room to account for the important role of others and material culture in the way we have come to know, leading to a simplistic view of cognition, interaction, intersubjectivity, and the ethical dimension; it eliminates the crucial role of social institutions and the values and tensions they convey. And it de-historicises knowledge (see, e.g., Campbell, 2002; Lerman, 1996b; Otte, 1998; Roth, 2011; Valero, 2004, 2009; Zevenbergen, 1996).

As we shall see in the following subsection, there are other ways to consider knowledge and the students' relationship to it.

2.2 *Sociocultural Approaches*

Sociocultural and constructivist approaches diverge greatly. They converge, however, in their opposition to transmissive pedagogy and their emphasis on the importance of the students' involvement in learning.[2] In sociocultural and constructivist approaches, to conceive of learning as the transmission and reception of knowledge is a type of behaviourism. Dogs learn how to successfully respond to certain stimuli; mice learn how to get out of a maze through specific *inputs*. The human mind, on the contrary, is much more complex; the behaviourist model of stimulus-response is definitely insufficient. In one now famous sentence, Vygotsky and Luria maintained that material and ideational culture mediates human behaviour and suggested replacing the stimulus response segment (S-R) with a triangle (Figure 5) that, in spite of its apparent simplicity, adds an inconceivable layer of complexity to the study of the human mind.

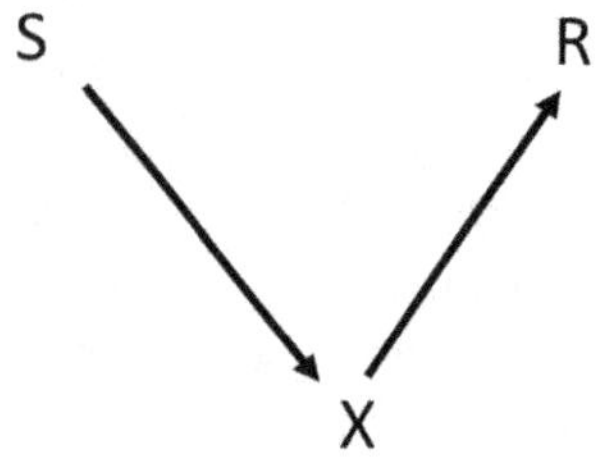

FIGURE 5
Vygotsky's famous triangle. External signs and other components of material and spiritual culture, X, alter the human mind's natural history (Vygostky & Luria, 1994, p. 144)

As we saw in the last chapter, humans carry out operations through signs. The important point is that the use of signs fundamentally alters the way we think and act. Vygotsky and Luria said: "With the transition to sign operations we not only proceed to psychological processes of the highest complexity, but in fact leave the field of the psyche's natural history and enter the domain of the historical formation of behaviour" (Vygotsky & Luria, 1994, p. 144).

However, if knowledge is neither something that one subjectively constructs nor something that one transmits, what is it? What is the difference with knowing? In the following section I present a cultural-historical concept of knowledge and knowing. The idea is to consider *knowledge* (for example, knowledge to solve linear equations or knowledge to think algebraically) not as a determined or delimited object but as a *generative capacity* to *do* something *or think* in a certain manner. Knowledge presents itself, thus, like a source of *empowerment*. In this context, knowledge does not exist in the individuals' heads. It is not a psychological or cognitive entity, but a *historical-cultural* one. Knowledge is a historical archetype of collective actions. Knowledge exists in culture and continuously emerges and changes through human activity. Indeed, through their activity, individuals trigger knowledge and put it in motion, thus making certain kinds of action and reflection concrete. When actuated, knowledge *shows* itself, and *materialises* into something perceptible, sensible, concrete. The *materialisation* of knowledge is what we shall call *knowing*. In other words, knowing is an *embodiment* of knowledge. In what follows, we shall see these ideas in more detail.

2.3 *Knowledge as Cultural-Historical Labour*

In the theory of objectification, knowledge is understood as a *general entity*. *Knowledge is a system of archetypes of thinking, action, and reflection constituted historically and culturally out of material, embodied, and sensible collective labour.*[3]

Let us imagine a rural community that, in the course of time, has produced ways of thinking, reflecting, and doing things—for example, how to sow the earth, how to think about space, quantity, time, etc. These ways of thinking, doing, and reflecting upon things are general archetypes that constitute the knowledge of the culture. Let us imagine now a culture based on capitalist ways of commercial production. This culture might produce a different knowledge—e.g., different archetypes of thinking, action, and reflection to deal with questions of selling and buying, interest rate, financialisation, etc. Knowledge changes from one culture to another and from one historical period to another.

Now, ontologically speaking, each one of us was born into a culture; in front of each one of us there was not only a world populated by concrete objects but also a world populated by ideas, by knowledge—cultural knowledge. For a baby born at this moment in the rural culture I mentioned above, there would be, in front of her, archetypes of thinking about the world. These archetypes would appear to her as potentialities for action and reflection. Another baby, born into the capitalist culture mentioned above, will find other potentialities. The central problem that I have to tackle in the next chapter is the problem of learning: How does the child learn about those archetypes that constitute the knowledge of her culture? For the time being, we need to continue exploring our concept of knowledge.

Aristotle used the term *potentiality* (δύναμις, dunamis) to refer to a capacity to do something (1984; see *Metaphysics*, 1048a). According to Aristotle, potentiality (δύναμις) is synonymous with power. Living beings and mechanisms have potentiality. A musical instrument, for example, has the ability to produce sounds. A fish has the ability to move in the water. *Actuality* (ἐνέργεια, energía), by contrast, is the concrete happening of that which, before being put in motion, before being actualised, was potentiality. That is why Sachs (2015, p. 3) says that actuality is being-at-work, something that happens before us, like the clear, concrete sound emitted by an instrument or the specific route followed by a fish in the water.

Potentiality is something undefined, without form, like sound before it is produced or like the capacity of the fish before it travels in the water; something purely potential that, through movement, becomes materialised (mat[t]er-ialised) or actualised (act-ualised: transformed by an act-ion).

However, the potentiality that living beings and mechanisms enjoy can be natural or acquired. The fish is biologically equipped to move in the water.

Other potentialities or capacities are acquired (Aristotle, 1984, *Metaphysics,* 1048a). This is the case of knowledge (see Hegel (2009), who precisely mentions mathematical knowledge as an instance of potentiality). Ontogenetically speaking, knowledge is this: generative capacity, potentiality. Algebraic knowledge, for example, is potentiality embedded in the culture: capacities that are offered to individuals to think, reflect, pose, and solve problems in a specific way.

It would be a mistake, though, to understand the idea of knowledge that I am outlining here as if knowledge were a kind of Platonic form. The fact that when each one of us was born and was confronted with a series of scientific, ethical, aesthetic, legal, and other forms of knowledge/thinking already established historically and culturally, does not mean that those forms of knowledge are Platonic forms. Platonic forms are universal and timeless, independent of human labour. On the contrary, knowledge, as I am articulating it in this book, is a dynamic entity: a cultural production, produced by concrete people through their labour—through their actions, their reflections, their joys, their suffering, and their hopes.

The following example may shed some light on the production of knowledge. It is about nut-cracking in chimpanzees.

Nut-cracking in chimpanzees is not an obvious process. As primatologists point out, nut-cracking includes the following steps:

FIGURE 6 Yo, the middle chimpanzee, breaks a nut while the young chimpanzees attentively follow the process (from Matsuzawa, Biro, Humle, Inoue-Nakamura, Tonooka, & Yamakoshi, 2001, p. 570)

1. The chimpanzee picks up a nut,
2. Puts it on a particular surface: an anvil stone,
3. Picks up and holds another stone (the hammer stone),
4. Hits the nut on the anvil stone with the hammer stone, and
5. Eats the heart of the broken nut (see Figure 6).

Studies conducted in the wild suggest that it takes from three to seven years for the young chimpanzees to learn the nut-cracking process. The youngsters do not necessarily start by using the hammer stone and the anvil stone. Proper attention to the objects, their selection (size, hardness, etc.), and the spatial and temporal coordination of the three objects (nuts, anvil, and hammer) is a long process. The six-month-old chimpanzees often manipulate only one object (a nut or a stone). They are able to pick a nut and step on it. As they mature, they may proceed to use the three objects, but not in the correct sequence, resulting in failed attempts. A key aspect of this process is the appearance of appropriate skills; for example, applying "sufficient pressure on a nut shell to break it" (Hirata, Morimura, & Houki, 2009, p. 98).

The chimpanzees learn to crack nuts as a social process. The young ones, who normally stay with their mother until the age of four or five years, attentively watch how their mother breaks nuts and then try to do it themselves, seemingly without understanding the goal of the process.[4]

Not all groups of chimpanzees crack nuts, and among the groups that do, not all crack the same kind of nuts. Primatologists believe that the ability to crack nuts developed somewhere in West Africa and was disseminated socially from one generation to the next. The practice of cracking nuts spread to neighbouring groups as a result of chimpanzee migration (Hirata et al., 2009, p. 88; Matsuzawa et al., 2001, pp. 569–70).

Cracking nuts is an example of knowledge: the series of steps summarised above constitutes, indeed, a cultural archetype of actions. As an archetype, knowledge is something *general*: it cannot be reduced to *this* or *that* particular sequence of coordinated actions with *these or those* stones. Its genesis is to be found in the chimpanzees' sensuous life. It exists in the chimpanzees' culture. We can think of knowledge as an *ideal form* of actions, as opposed to the actions themselves. Knowledge as an ideal form is related to each one of its concrete instances or realisations, but is at the same time different from each one of them. Each of these concrete realisations keeps, in its materialisation, the generality of the ideal form that engenders it, but it does not coincide with the ideal form.

As previously mentioned, knowledge has nothing to do with Platonic forms. Instead of considering the nut-cracking community of chimpanzees that lives in the jungles of Mount Nimba in the Republic of Guinea as Platonic forms or Kantian *things-in-themselves*, we can consider it as an archetype of embodied and material reflection and action constituted culturally and historically. The "ideal form" to crack nuts should be understood as a general or archetype of doing things. Rather than being located in a realm of eternal ideas, this ideal form is encoded in cultural memory. Unlike the Platonic forms, whose nature is precisely to exist independently of what the different species do on earth, knowledge is an ideal form that results from concrete practice and that cannot exist if it is not realised in such practice.

The example of knowledge that I have just discussed has, it seems to me, a pedagogical virtue. It is a "simple" example as long as language or other complex semiotic systems do not intervene, as is generally the case with human knowledge. The nut-cracking knowledge is, like other knowledge produced by chimpanzees (e.g., washing potatoes with water or brushing off ants with the branches from shrubs), *kinesthetic*.

It would be worth it for us to stop now on an example of mathematical knowledge. Let us tackle the case of *pattern generalisation*.

Like many of my colleagues, in my classroom research, I have used pattern generalisation to introduce the students to algebra (see, e.g., Bednarz, Kieran, & Lee, 1996; Cai & Knuth, 2011; Kieran, 2018; Rivera, 2011). The fundamental idea is to present to the students geometric sequences or simple numbers (usually arithmetic sequences that can be expressed in a linear form: y=ax + b, like the problem of the Piggy Bank discussed in Chapter 1). Usually, we give the students a few terms (see Figure 7) and then ask them to come up with various ways to calculate the number of objects in "remote" terms (e.g., Terms 10, 25, 100).

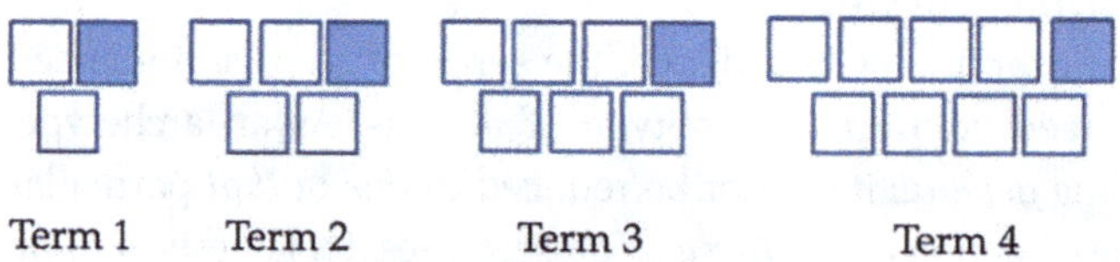

FIGURE 7 The first terms of a sequence that students in Grade 2 investigated in a lesson

In so doing, we hope that the students enter into a relationship with a *cultural-historical system of actions and reflections*: very specific ways (that we institutionally recognise as algebraic) of perceiving, reflecting, producing, and investigating sequences by way of symbols, diagrams, concrete objects, etc. This system of actions and reflections is mathematical knowledge—more precisely, algebraic knowledge.

Like all knowledge, algebraic knowledge was constituted historically. Indeed, the investigation of arithmetic sequences appeared in ancient civilisations, such as Mesopotamia, and was a very popular subject in Late Antiquity in neo-Pythagorean circles (Lawlor & Lawlor, 1979; Nicomachus of Gerasa, 1926; Tarán, 1969). The neo-Pythagoreans were particularly interested in polygonal numbers—numbers represented by the shape of a regular polygon.

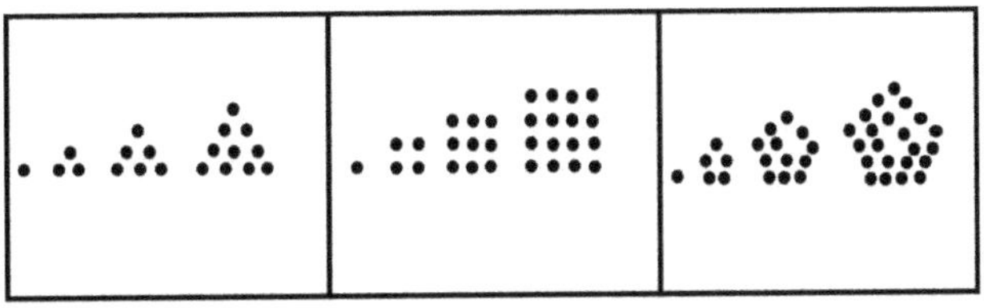

FIGURE 8 The first triangular, square, and pentagonal numbers

For example, the first triangular numbers are 1, 3, 6, 10; the first square numbers are 1, 4, 9, 16; and the first pentagonal numbers are 1, 5, 12, 22 (see Figure 8).

Supported by historical records, historians think that the first Greek investigations of sequences were made with little stones arranged in the form of figures (as suggested in Figure 8). In this way, the forms of action, reflection, and investigation of the first Greek research on sequences would have been based on perceptual arguments. It is also in this perceptual way that the first properties of even and odd numbers would have been established (Lefèvre, 1981).

According to historians, research into the properties of arithmetic sequences based not on perceptual arguments but on *deductive* arguments first appeared in Hypsikles's text known as *Anaphorikos* (see Radford, 2006).[5] Proposition 1 is stated as follows:

> If any number of terms is considered such that <starting from the greatest> every two successive ones have the same difference, [the terms] being even in number, then, the difference between [the sum of] half the number of terms [starting from the greatest], and [the sum of] the remaining ones, is equal to the multiple of the common difference by the square of half the number of terms. (Manitius, 1888, p. 2)

In modern symbolism, the proposition states that if a number of 2n terms, $a_1, a_2, \ldots, a_{2n}$, are such that $a_1 > a_2 > \ldots > a_{2n}$, and $a_i - a_{i+1} = d$ for i = 1, 2, …, 2n – 1, then:

$$(a_1 + a_2 + \ldots + a_n) - (a_{n+1} + a_{n+2} + \ldots + a_{2n}) = n^2 \times d.$$

Hypsikles's proposition states a *property* of what we now call an arithmetic sequence. Diophantus (ca. 250 AD), in his concise text, *On Polygonal Numbers* (see Ver Eecke, 1959), presents a formula for calculating any polygonal number, *Sn*, when the side n of the polygonal number and the angle a are known. The formula is:

$$S_n = \frac{[(2n-1)(a-2)+2]^2-(a-4)^2}{8(a-2)}$$

Let us suppose that we want to calculate the third term of the pentagonal numbers. In this case, $n = 3$ and $a = 5$.

$$S_n = \frac{[(2\times 3-1)(5-2)+2]^2-(5-4)^2}{8(5-2)}$$

which yields $S_3 = 12$.

Naturally, Diophantus did not express this formula using modern symbolism. What he did tell us is:

> Take twice the side of the polygonal number; from this subtract one unit; multiply the result by the number of angles minus 2; then add 2 units. Take the square of the resulting number. From this, subtract the square of the number of angles minus 4. Divide the result by 8 times the number of angles minus 2 units. This gives us the polygonal number we are looking for. (Based on the translation of Ver Eecke, 1959, pp. 290–291)

To summarise, this example suggests that a first form of perceiving, producing, reflecting, and investigating numerical sequences based on pebbles arranged in certain ways was transformed into a different form based on the investigation of analytical-deductive theoretical properties of numerical sequences. But the example is more than a historical exercise: it allows us to realise the ontological constitution of knowledge as understood in the theory of objectification. In its ontological constitution (that is, in its intimate *nature*), knowledge is understood as a systemic, dynamic, and fluid "whole" composed of individual components and their connections. Knowledge is considered as an entity "multivariously divided within itself, rich in determinations, and historically formed. It is like, not a separate isolated atom, but a living organism, a socioeconomic structure, or similar formation" (Ilyenkov, cited in Bakhurst, 1991, p. 138). To say that knowledge is a dynamic system signifies that the study and investigation of knowledge is the study of its *movement*, of its continuous transformation, and of the movement of its parts and interconnections.

In this line of thinking, the early Greek mathematical knowledge about sequences as practised by the early Pythagoreans was *refined* and *transformed* into something more *specific* and *determined*: something that now reveals new *interconnections* between the polygonal number Sn, the side n, and the number of angles a that, as historical testimonies suggest, had not been previously observed.

Dialectical materialism has a specific term to explain the movement and transformation of knowledge. New knowledge is, with respect to the previous, more *concrete*. Clearly, the meaning of "concrete" in dialectical materialism has a different meaning from the empiricist one. In the latter, the concrete is seen as something that, as we touch it, leads us to form abstract ideas. In empiricist thought (see, e.g., Locke, 1825), each idea supposedly embodies properties of empirical objects, "sharply and fixedly separated from each other" (Inwood, 1992, p. 30). In the dialectical materialist view adopted in the theory of objectification, the concrete is rather related to fluid links between its components; it is related to the *movement of knowledge* and to new *understandings* of the world (Ilyenkov, 1982).

This process of movement, transformation, and refinement of knowledge occurs through a process of *determinations* that follow one another and in which new connections appear. New *connections* of knowledge do not merely replace the old connections; the new connections convey, in a condensed manner and with tensions, the meanings of previous theoretical formations. The emergence of alphanumeric symbolism with Vieta and Descartes reveals new links and enables deeper distinctions between classes of numbers (Filloy & Rojano, 1984; Rojano, 1996). In this historic process, the previous meanings become subsumed in the new meanings. For example, in the work of Renaissance mathematician and painter Piero de la Francesca, the unknown in algebraic problems is named "the thing." The unknown is conceptualised in association with the *side* of a square, so three unknowns are written as $\overline{3}$; the side of the square appears on top of the number, as a sign that signifies *iconically* (Radford, 2006). It will take more than one century to reach the modern symbolisation, $3x$ (Høyrup, 2008). This example shows how the previous meanings are sedimented and become subsumed in the symbolic expression. Between the historical moment in which the 3 unknowns were thought of as $\overline{3}$, to the moment they were thought of as $3x$, there has been a change not only in mathematics symbolism but also in mathematical knowledge. In the contemporary school, children do not find $\overline{3}$ in its original geometric Renaissance conceptualisation. They find a knowledge that has undergone a process of transformations that makes it more definite, more concrete. As Marx said, "The concrete is concrete because it is the concentration of many determinations, hence unity of the diverse. [The concrete] appears in the process of thinking, therefore, as a process of concentration, as a result, not as a point of departure" (1973, p. 101). It is only within an empiricist perspective that things appear backwards, where the concrete (understood as sensible) is taken as the starting point (but then, we get into the problems that surround Feuerbach and his "sensuous certainty").

The problem that I am dealing with here is not simply a problem of terminology: it is not a question of different meanings, such as in empiricism where the concrete relates to the sensible and perceptible, while in dialectical materialism the concrete refers to the movement of knowledge. Behind this difference is an irreducible contrast between two concepts of knowledge. As Bakhurst notes, one of the main problems with the empiricist concept of the concrete is that it assumes that ideas are generated through the individual's sensible experience of the world and forgets that "for each individual, the starting point of cognition is not unprocessed sense experience but a conception of the world *inherited* 'ready made' from the community of which he or she is a member" (Bakhurst, 1991, p. 150, original emphasis).

Within the concept of knowledge that I present in this chapter, the production and transformation of knowledge must not be considered as something that occurs as if it were achieved by an invisible hand or by an alleged intrinsic logic of knowledge development (as rationalism does). The evolution of knowledge is conceived not as a natural phenomenon but as a *cultural* one. Just like capital can only be understood as the historical realisation of abstract concepts such as the division of labour, currency, and value, among others, mathematical knowledge can only be understood as a realisation of previous abstract forms of mathematical thinking and action. So, when I speak of determinations that follow each other, revealing in that process new connections in the object of study, I do not want to say that, from a phylogenetic point of view, those connections were *already there*, waiting to be discovered. The connections are the result of human labour; they are not absolute determinations given outside culture, politics, and economy (see Chapter 8). They appear within the limits and possibilities of the "sum of productive forces, capital funds and social forms of intercourse, which every individual and every generation finds in existence as something given" (Marx, 1998, p. 62). The production and transformation of knowledge occurs, in other words, within the limits and possibilities of that which is the real basis of human existence at a specific time and place.

It is precisely why, in each historical moment of a culture, knowledge contains certain parts and certain connections between its parts and no others, that such knowledge offers individuals certain *generative capacities* of action, reflection, and thought specific to this historical moment. Vygotsky makes this point in his text on imagination and creativity when he says

> Every inventor, even a genius, is also a product of his [sic] time and his environment. His creations arise from needs that were created before him and rest on capacities that also exist outside of him ... No invention or scientific discovery can occur before the material and psychological conditions necessary for it to occur have appeared. Creation is a historical, cumulative process where every succeeding manifestation was determined by the preceding one. (Vygotsky, 2004, p. 30)

The capacities Vygotsky is talking about that exist outside the individual are what I have termed knowledge.

My example of knowledge about arithmetic sequences does not contain anything special. Similar examples could be given about any topic in mathematics. The important thing is, therefore, that mathematical knowledge has been created through a lengthy process of refinements and concretions (natural language, alphanumeric symbolism, graphs, etc.) and has been codified in memory and cultural practices, and is now present in many curricula in many parts of the world. It is this knowledge that the students encounter in school and that would lead them to see that Term 100 of the sequence shown in Figure 7, for example, has $1 + 2 \times 100$ squares.

In the next section I suggest a definition of *knowing*.

3 Knowing

In its *ontological* constitution, that is to say, in its *constitutive* nature, knowledge is a dynamic, and fluid "whole" or system composed of individual components and their connections. In its *epistemological* constitution, that is, as a *generative capacity*, knowledge is a cultural-historical way of doing, thinking, and reflecting. In this line of thought, knowing can be defined as the *actualisation or materialisation* of knowledge.

When I say that knowing is the actualisation or materialisation of something that already exists, I run a great risk of being misinterpreted. Knowing can appear as a simple repetition. Of course, that is not true. Knowledge is potentiality. Knowing is a materialisation of knowledge. It is something that we can perceive, notice, feel: something that can be grasped by human consciousness. So, when I suggest that knowing is an actualisation or materialisation of knowledge, we need to distinguish between:

1. Knowledge as a *general* entity.
2. The *process* through which knowledge is actualised or materialised.
3. Knowing as *actualisation* or *materialisation* of knowledge.

Let us look carefully at these three points.

3.1 *Knowledge as a General Entity*

To state that knowledge is something *general* means that knowledge cannot be reduced to any of its materialisations or actualisations, like in the chimpanzees' example. It is to affirm what we have said previously: that knowledge is a generative capacity of action and thought: the possibility and capacity to break this or that nut; the possibility of finding a property of arithmetic sequences or the term 100 in a given sequence. This possibility is a potentiality that "has not yet come into existence" (Hegel, 2001, p. 36), like the potentiality of a musical instrument. In order to emerge into existence, knowledge must materialise through a *process of actualisation.*

3.2 *The Process of Actualisation of Knowledge*

The term actualisation evokes a temporary nature. There is already something there, but it is just potentiality, and in order to emerge into existence, it must be put into motion and *appear*: it has to turn into something *actual*; it has to be *actualised.*

The name of this *process* through which knowledge is actualised is *activity*: through activity knowledge acquires a sensible determination (or content) and as a result knowledge is actualised. "It is only by this activity that [the general] as well as [its] abstract characteristics generally, are realised, actualised; for of themselves [the general and its abstract characteristics] are powerless" (Hegel, 2001, p. 36). In other words, the general (knowledge) does not have the power to appear by itself. It is this Hegelian concept of knowledge that Vygotsky (1987) referenced when he asserted that any conceptual phenomenon must be studied as a process in motion and change.

However, *activity* is not a simple channel through which knowledge makes its appearance. It has already been mentioned in Chapter 2 that this concept of activity (Tätigkeit) is a dynamic, complex system that includes the energy that individuals put in motion, an energy that is configured according to the individuals' own motives and biography, as well as to the cognitive and material possibilities offered by language, signs, and artefacts. We will return later to this concept of activity that is central in dialectical materialism and the theory of objectification. For the moment, let us pause on the idea of actualisation.

3.3 *Actualisation*

Knowing is the concrete conceptual content through which knowledge is embodied, or materialised, or actualised. However, its conceptual, concrete content appears and can appear only through human activity. This activity actualises knowledge, brings it to life—like the activity of playing a violin brings musical notes to life. This means that between knowledge and knowing lies activity. In other words, *knowing is the result of a mediation*. There is no such thing as unmediated knowing: all knowing is mediated, and what mediates it is an activity.[6]

However, we should not think that the mediating activity is something *given*, like a bridge that joins two entities already given. Activity is an entity in *dialectical* relation to knowledge and knowing. Indeed, activity is something that unfolds in time and space, that moves knowledge and, while moving it, materialises it and converts it into knowing. *At the same time*, in the unfolding of the activity, the same materialisation *affects* the activity. There is a close dialectical relationship between the activity and that which is appearing. In the end, the meaning of this mediation is the following: *knowing bears the footprint or the impression of the activity that mediates it* (Ilyenkov, 1977). The activity demarcates the manner in which knowledge manifests itself in knowing. In even simpler terms, the way we come to know something (for example, how to solve equations) is consubstantial of the specificities of the activity through which knowledge reveals sensibly to human consciousness.

Let me summarise the relationship between knowledge, activity, and knowing in the following way. Knowledge is a *general* entity. We humans cannot sense nor think of knowledge (algebraic, geometric, etc.) *as such*, as *general*. A case in point is the remark made by the mathematician and didactician Laurent Vivier (2020) in a recent seminar in Paris. Vivier remarked that we are unable to deal with the notion of tangent *in general*. A tangent *as such* is not thinkable. We need a context—e.g., a functional, cinematic or geometrical context. To become an object of thought and consciousness, knowledge indeed has to be put into motion. What puts the notion of tangent or knowledge more generally into motion is human sensuous-intellectual-material contextual activity (Tätigkeit). By being put into motion, knowledge acquires contextual *determinations* and is transformed into a singular form. In our terminology, this singular sensible form is *knowing*.

4 The Piggy Bank Example

Let us return to the Piggy Bank example that we discussed in Chapter 1 and consider it in light of the previous ideas.

Before beginning the lesson, algebraic knowledge exists as a *generative capacity* for action and thought about the problem of the sequence of savings. Algebraic knowledge is the result of a historical process and exists in the students' culture. The teacher or another individual already familiar with algebraic knowledge can effectively solve the problem: he can implement a series of actions leading to find, by culturally recognised algebraic procedures, the amount of money saved in Weeks 10, 25, etc. However, until that morning, this knowledge remained beyond our Grade 4 students' recognition. Indeed, as discussed in Chapter 1, the students resorted to arithmetic procedures and concepts (see Figure 2). The teacher organised the classroom activity in such a way that, working collaboratively with the students, algebraic knowledge could progressively acquire determinations and be transformed into a singular sensible form. This organisation includes the choice of questions and the order in which they appear on the worksheet. It also includes the forms of human cooperation (e.g., the interactional organisation of the classroom space, small group work, discussions, etc.). The sought-after algebraic knowledge that in its generality still escapes our students' consciousness and thought is set in motion by the classroom activity. Through the joint labour of the teacher and students, algebraic knowledge begins to *move* and *show itself*; knowledge begins to take shape and materialise. And that which materialises is knowing. Knowing is this *common work* mentioned in Chapter 2: sensible collective work of thoughts and actions carried out with the body, artefacts, and speech that are so intertwined that there is no possible distinction between them. Algebraic knowing is that which appears—or begins to appear—and insinuates itself in the famous line 29 and the lines that follow. Algebraic knowing appears through language, perception, the body posture of the students and of the teacher, the bingo chips, the colours, etc.:

29. Mrs. Giroux: (*Trying to make noticeable to the students the co-variational structure*). What do you remark about Week 5 (*She shows the glass corresponding to Week 5*) and (*Pointing to the red bingo chips; See Figure 3, Picture 4 in Chapter 1*) the number of bingo chips? (*Making the same actions*) The fourth week and the number of bingo chips?
30. Albert: It's always twice ...
31. Mrs. Giroux: (*Repeating*) It's always twice.

32. Krysta: It's the double of what you ... No! (*She watches the artefacts intensely for a while*) I am confused!
33. Albert: Yeah! It's twice, look! (*Counting the red chips*) 1 + 1, 2; 2 + 2, 4; 3 + 3, 6; 4 + 4
34. Krysta: (*Interrupting*) 8

Knowing is all *that* which begins to materialise in those lines. It is the materialisation of a generative capacity to understand, think, reflect, and act on the problem in co-variational terms; it is the generative capacity *in action*, transformed and materialised into something sensible through the mediation of joint labour.

Evidently, algebraic knowledge is not disclosed *totally*. That which is disclosed is a singular, developed form: that which happened in the classroom that morning. But that which happened and showed itself that morning will allow the teacher and the students to continue their adventure into mathematics knowledge. Later on, the discussion can focus on other problems or the same problems to try to achieve a deeper understanding. This is what occurred in this class, as we shall see in the next chapter. What occurred is something of a more general nature that has to do with the dialectics between knowledge and knowing and their development. It should not be a surprise hence to find this dialectics in the historical transformation of knowledge and knowing. In the next section I present an example to illustrate this idea.

5 The Dialectic between Knowledge and Knowing

Some Babylonian clay tablets show counting problems. A sign impressed on an argil tablet is the sign of a unit (Damerow, 1996; Høyrup, 2002; Jasim & Oates, 1986; Schmandt-Besserat, 1992). The earliest simplest numeric configurations were created by *iterating* a sign for the unit. These "numerals" had the purpose of facilitating the systematic calculation of additions and subtractions. Embedded in practical activities oriented to the solution of administrative and other societal problems, the constructive-additive representation of numbers was systematised and converted in a cultural way of counting; that is, *knowledge*—elementary arithmetic knowledge that, on the cognitive side, went hand in hand with the emergence of an elementary cognitive arithmetic additive structure. Although strings of iterated signs were useful to cope with some societal demands, counting large amounts of objects was not practical. The cultural forms of counting appeared in the concrete world and had to be expanded to count big collections of objects. So, later on we find the introduction of new signs to replace strings of sign-unit iterations and the ensuing rules of symbol-substitution, leading eventually to a positional numerical system. The expansion of knowledge was achieved through human activity. It was in human activity, indeed, that new forms of counting (counting by groups, finding signs for the new groups, etc.) appeared and underwent a new process of social systematisation, leading to new knowledge. Without the possibility of becoming actualised—that is, of being transformed into knowing—knowledge would remain general and without possibilities to be modified and expanded.

Figures 9a and 9b try to capture the relation between knowledge (*K*), activity (*A*), and knowing (*k*). From a phylogenetic point of view, at one point in the development of a culture, knowledge *K* (see Figure 9a) is set in motion by human activity (symbolised by the arrows) and actualised or materialised in a developed form (as knowing, *k*). Through the activity, which is always movement that is affected by *K* and the emerging *k*, the individuals can now refine, adjust, expand, and transform knowledge *K*, giving as a result a new knowledge *K'*. The new knowledge *K'*, converted into new potentiality, can, through the mediation of other activities (the arrows in Figure 9b) reveal or actualise itself in another knowing *k'* (see Figure 9b).

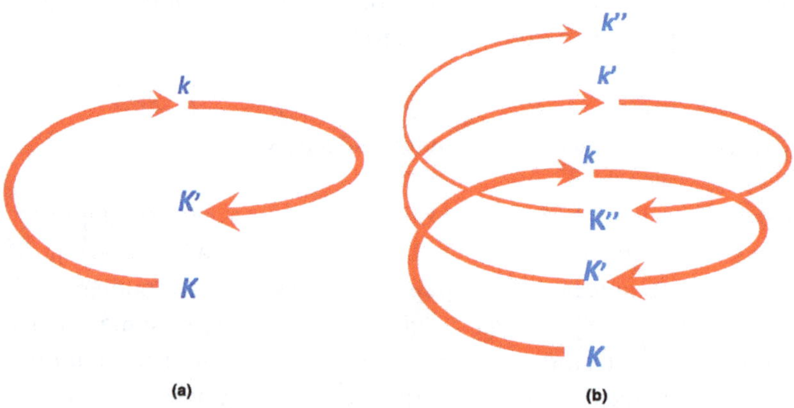

FIGURE 9 Through activity, knowledge (*K*) acquires a sensible content, and is materialised as knowing (*k*). New systematisations occur, leading to an expansion of knowledge. The resulting new knowledge (*K'*) leads through activity to new knowing (*k'*), etc.

Knowing is the concrete, conceptual content of knowledge: it is the manner in which potentiality is transformed into actuality (Maybee, 2009). In the same way as the sound of a violin is the aural content of its potentiality,

FIGURE 10 Grade 6 students materialising a cultural-historical algebraic way of thinking (knowledge) through a classroom activity. The materialisation consists of reflections and actions on the given equation

mathematical knowing is the theoretical content of mathematical knowledge. The way in which knowledge acquires its content depends on the activity that sets it into motion.[7]

When the Grade 4 students we mentioned in Chapter 1 reached Grade 6, they solved the equation shown in the first line of Figure 10 (i.e., the equation $5n + 3 = 3n + 19$). To solve this equation, the students materialised a cultural form of action and reflection (knowledge). The materialisation of knowledge occurred through a particular classroom activity that included the use of symbols and artefacts, language, gestures, body posture, etc. Out of this activity appeared knowing: an intellectual-material manner of doing/reflecting on what it takes to solve the mentioned equation and that is at the same time sensual and sensible, material and ideal.

Knowing as the materialisation of knowledge is always a *singular* (i.e., *unique*) *developed form*. The actualisation is not able to capture knowledge in its totality. In the example of Figure 10, in the materialisation of knowledge, subtractions of unknowns and positive numbers appear. By contrast, there are no reflections on fractions or negative coefficients. In that sense, knowing is always a *deficit*. But, at the same time, knowing is *excess*: given its sensible material nature, knowing exceeds potentiality and makes way to set out new problems, to create new lines of reflection and research. Therefore, in embodying knowledge, knowing *confirms* and *affirms* knowledge; and, at the same time, as excess, *denies* it. The dialectic principle of knowledge and knowing is not a principle of identity: it is a principle of *difference*.

In the section that follows I continue discussing the dialectic principle between knowledge and knowing and present, in some detail, an idea that derives from there: the idea of mathematics as something that is at the same time ideal, sensible, and material.

6 Mathematics as an Entity at the Same Time *Ideal, Sensible, and Material*

The dialectical concepts of knowledge and knowing introduced in the previous sections lead to a specific conception of mathematics and its nature. I had the opportunity to outline this conception some years ago, during a one-week Colloquium on Symbolic Cognition that Stephen Hegedus and the Kaput Center organised in a secluded and remote cottage-hotel in Vermont, in the USA.

Like many other participants of the Colloquium, I landed at Boston Airport. A taxi was waiting for me. As I got into the taxi, the driver let me know that we had to wait for another participant who was supposed to arrive anytime. "And who is it?" I asked. The driver consulted his log sheet and responded, "Mr. Colette Laborde." I corrected him and said, "*Mrs.* Colette Laborde." He checked his log sheet again and said emphatically, "*Mr.* Colette Laborde," and added "and he is travelling with skis."

While we circled the various doors of Boston Airport looking for a man with skis, I tried to convince the driver that there was no one called Mr. Colette Laborde attending the meeting; I drew on Toulmin's model of argumentation, backed up my claims with carefully chosen warrants, but failed miserably. Then I spotted Colette in front of one of the doors. I waved and she waved back. The taxi stopped and she got in. She was not bringing skis. The driver was surprised, called headquarters, and complained. Then we left. We drove for about 45 minutes and arrived at night at the cottage-hotel. Most of the participants had already arrived. The group was composed of about 15 researchers working on symbolic cognition and semiotics. They included Norma Presmeg, Adalira Sáenz-Ludlow, Nathalie Sinclair, Hyman Bass, David Tall, Gerald Goldin, and Gari Davis.

As part of the program, in the afternoon of one of the first days of the Colloquium, we had a special guest: neuroanthropologist Terrence Deacon, author of *The Symbolic Species* (Deacon, 1997). Deacon got out of a car, went directly to the main room, unpacked some notes, took two long breaths, and gave a remarkable talk about chimpanzees' language learning.

The following day I was having breakfast with some mathematicians and mathematics educators at a table by the window. We could see the beautiful, totally white, landscape. It had been snowing without interruption for days. It was January and it was cold. I do not remember what brought us to discuss the nature of mathematics. Maybe it was a good night's rest, or the interesting discussion on symbols the day before after Deacon's talk, or both. I ventured to mention that mathematics could not be equated to symbols. Mathematics, I argued, is not comprised of the symbols on the pages of a book. The symbols on the pages of a book are exactly that—symbols, or marks, to put it more bluntly.

To support my claim, I resorted to music. In the same way that there is no music in a score sheet, there is no mathematics in the pages of a textbook. Music is what we *hear* when people play instruments. Music is a sensuous aural entity.

Mathematics is also a sensuous entity: it appears much in the same way as music when an orchestra plays, say, a symphony. Like music, mathematics is something that appears as students and teachers engage in classroom activity. What appears in the mathematics classroom is not exactly an aural phenomenon or a visual or tactile or olfactory one. Yet, something appears (and perhaps is something that is all of that: visual, tactile, olfactory, aural, material, artefactual, gestural, and kinesthetic) and, being all of that, becomes an object of consciousness and thought. School mathematics, in this materialist and phenomenological line of thought, is what is made sensible through the teachers' and students' activity.

To continue with the Vermont metaphor, we need to make some distinctions. I am not saying that mathematics and the activity that produces it are the same, much as the music that is being heard cannot be confounded with the orchestra's activity. Yet, both are deeply intertwined: the activity and what the activity produces. We cannot extract one from the other: we cannot extract the mathematics from the classroom activity, as we cannot extract the orchestra's activity from what we hear. As the activity unfolds, mathematics appears—much as, for example, a symphony appears as the orchestra's activity unfolds. Unfolding and appearing have to be understood here in a dialectical relationship. This is exactly the dialectical relationship I mentioned in Section 3 above. The unfolding affects, moves, and transforms the appearing, and the appearing affects, moves, and transforms the unfolding.

But things do not merely happen or appear out of the blue. The sound that is produced by a violin, for example, has, as mentioned previously, its source in the instrument. The sound may or may not be produced. And *if* it is produced, it can be produced in countless ways. It is both contingent and historically bounded. In being produced, the sound materialises or actualises that which was potentiality or pure possibility. If we return to Hegel's terminology and related conceptualisations, we can consider both mathematics and music as sensuous evolved forms of something that before being materialised and coming into sensible existence, was *general*. The general is formless. It belongs to the realm of potentiality. Yet (again!) it is not a Platonic Form. The realm of the potential belongs to an always changing immaterial sphere of culture that is *intertwined* with the material world of objects and human actions. This immaterial sphere of culture is part of what Marx (1988, p. 75) called the "inorganic" realm of nature and it is also part of the conditions out of which human existence is crafted. This sphere cannot be sensed by we humans through our culturally and historically evolved senses and sensations. Can we sense or perceive or touch the Pythagorean theorem *as such*? We cannot. Can we hear Beethoven's 7th Symphony *as such*? We cannot. To be grasped by consciousness and thought, to be felt, the general (mathematics *as general*; music *as general*) has to be set into motion to transform it into something sensible, and appear. Its appearance is what in Hegel's terminology is called a *singular* (and that, in the case of mathematics, we have called *knowing*). The singular is the appearance of the general through the mediation of human activity.

To make the previous idea clearer, let me turn to Beethoven's music—which Adorno (1998, p. 224) considered "bourgeois music at its very height." Let me turn to Beethoven's 7th Symphony. As we know, Beethoven's 7th Symphony has four movements: Poco sostenuto—Vivace, Allegretto, Scherzo, and Allegro. Table 2 presents the duration of the symphony as conducted by two orchestra directors who are considered to be among the best 20th century Beethoven specialists: Herbert von Karajan and Leonard Bernstein.

Bernstein's recording is 12% longer than Karajan's. And I think that the reader would agree with me that 12% is a lot. It is not an insignificant difference. Which one is the true 7th Symphony? Neither of them. The 7th Symphony as such is a *general*. Bernstein's and Karajan's recordings are materialisations of

TABLE 2 Total time of two famous recordings of Beethoven's 7th Symphony

	Herbert von Karajan (Berlin Philharmonic Orchestra, 1963 recording)	**Leonard Bernstein (New York Philharmonic Orchestra, 1958 recording)**
Poco sostenuto – Vivace	11:25	12:27
Allegretto	8:02	9:44
Scherzo	7:50	8:23
Allegro	6:37	7:27
Total time	33 min 54 s	38 min 01 s

this general; that is, they are singulars, or, in other words, they are appearances of the general: what we have termed *knowing*.

As all generals, the 7th Symphony is inscribed within a certain cultural tradition. It belongs to a musical tradition of leisurely symphony prologues, a romantic paradigm, an increasing focus on rhythm and the smart use of available musical artefacts (e.g., metronomes for measuring tempo), among others (Marx & Burnham, 1997; Will, 2004). Rather than existing in itself and by itself, the general is to be found in culture and history.

I would like to summarise these ideas by saying that the singular (knowing) is the appearing of the general (i.e., knowledge: mathematical knowledge, musical knowledge, etc.).[8] In other words, the singular is the coming into existence of the general as an evolved ontological form transformed under the force of an activity (in the sense of Tätigkeit/deyatel'nost', as defined in Chapter 2). That this activity is not merely an Aktivität or aktivnost' is shown by the fact that the Aktivität or aktivnost' would produce something else, something like a 7th Symphony interpreted by programmed artefacts and mechanical devices only. Such appearing would in fact lack exactly that which makes Tätigkeit/deyatel'nost' what it is, namely *human, natural life*.

Let us come back now to mathematics and mathematics classroom activity. What human activity produces is mathematics as a sensible phenomenon—a singular evolved form of something general that before being set into motion by the classroom activity was potentiality, pure possibility. The activity that happens in the classroom can be "good" or "bad." The bad activity is precisely that which looks pretty much like the mechanical one of my example, where people do not really connect, where they do not work together. They simply do things, as in Aktivität or aktivnost'. It is lifeless activity (D'Amore, 2015). As it will become clear in the next chapters, in the theory of objectification we are interested in nurturing activities where students and teachers engage, debate, agree and disagree, where they object and find a place for subversion, and where students could show solidarity, responsibility, and care for each other.

7 Synthesis

In this chapter I dealt with the concepts of knowledge and knowing in the theory of objectification. I suggested that knowledge is a general entity—a system of forms or *archetypes of thinking, action, and reflection constituted historically and culturally out of material, embodied, and sensible collective labour*. I drew on Aristotle's and Hegel's distinction of potentiality and actuality to distinguish between knowledge and knowing, where the latter is considered as a developed sensible version of the former. The starting point of my theorisation was the rather banal fact that, when we are born, knowledge as a system of forms (always changing, always being transformed) exists in our culture (how to count, how to think of and distinguish shapes, etc.). We encounter knowledge and participate in its transformation as we engage in cultural activities.

Now, discussions in seminars and other contexts have shown me that sometimes the idea of knowledge that I am articulating here can be interpreted as if what is proposed in the theory of objectification is an idea of knowledge of a platonic nature. This misconstruction explains the effort I have made in this chapter to dispel misunderstandings. But there is another source of misunderstanding—to think that knowledge in this theory is something that individuals *internalise*. In my previous work I have already mentioned some of the shortcomings of the idea of knowledge as internalisation (see, e.g., Radford, 2018b; I come back to this point in Chapter 4). Yet, Roth (2020) mistakenly says

> Mathematical forms tend to be theorized as signs (things) that refer to something else (e.g., "meaning"). The apotheosis of this approach exists in/as the Theory of Knowledge Objectification (Radford 2013), whereby internalized knowledge manifests itself in external mathematical forms that can be shared within groups of individuals. These mathematical forms thus refer to something existing on the inside of the person that otherwise is inaccessible; the forms are signs pointing to something else. (Roth, 2020, p. 91)

Roth's mistake consists of understanding our idea of knowledge as a problem of semiotic reference. He misconstrues the relationship between knowledge and knowing as if one *refers* to the other (as in x and f(x)). In the theory of objectification, the relationship between knowledge and knowing is *dialectical* in a materialist sense. First, mathematical knowledge is a system of *forms*, not signs. Knowledge is a *system* of forms of action and thinking (e.g., adding numbers in the base-10 system as we do or in the base 60 as Babylonian scribes did); its genesis is to be found in sensuous, concrete, cultural, historical, practical activity. Second, it is this practical activity that brings knowledge into life. But what appears (i.e., knowing) is not an f(x) of an x. There is no such a thing as f. This is a terribly static interpretation of the dynamic dialectic relationship between knowledge and knowing. As I insisted in this chapter, knowing is an *excess* vis-à-vis knowledge. What this excess means is that knowing is not determined by knowledge; knowing is *overdetermined* by knowledge *and the whole cultural-historical context*. Without this overdetermination, we would be repeating things like marionettes. Knowing *negates* and *subverts* knowledge. In concrete activity, we are always going beyond the given. This is how Descartes could imagine analytic geometry and went beyond Euclidean geometry. In Chapter 8 I come back to the question of overdetermination when I discuss the concept of culture and how cultural knowledge affects the way we think and do things.

In the last section of this chapter I made an excursion into the realm of music and musical knowledge to try to better convey the idea of mathematical knowledge as something that is at the same time ideal, sensible, and material. Mathematical knowledge, as all knowledge, is coming continuously into reality through practical activity; what appears is form-and-content, the ideal-sensible in its materiality.

Acknowledgement

This is a revised and expanded version of Radford, L. (2013). Three key concepts of the theory of objectification: Knowledge, Knowing, and Learning. *Journal of Research in Mathematics Education*, 2(1), 7–44.

Notes

1 One exception is the theory of didactic situations (Brousseau, 2002), which uses the terms *savoir* and *connaissance*.

2 This particular convergence has led some mathematics educators to locate Vygotsky among the constructivists. Vygotsky appears as a social constructivist, which seems to me to ignore the important fact that Vygotsky's work and constructivism draw on philosophical paradigms that are completely different. Erasing this difference would lead us to see Vygotsky as a Kantian and Piaget as a Marxist.

3 To characterise knowledge, in previous works I have used the term "set." I wish to thank Rodolfo Vergel for suggesting the use of the term "system," which better emphasises the idea of interrelated elements or units in movement.

4 For instance, they play with the stones used by adults in cracking nuts; see http://www.youtube.com/watch?v=5Cp7_In7f88

5 Hypsikles lived in Alexandria. Historians do not know much about his life, but they think he lived between the 2nd century B.C. and the 2nd century A.D.

6 "Activity," Andy Blunden says, "is simultaneously subjective and objective; activity is the 'middle term' mediating between subject and object" (2010, p. 98).

7 In light of Figure 9, it is worth mentioning again (see Chapter 2) the theoretical differences between, on the one hand, the concept of activity at the heart of the theory of objectification and, on the other hand, the concept of activity adopted by some contemporary theories of action. In the theory of objectification, activity (Tätigkeit) is what sets cultural-historical knowledge into motion, transforms it into knowing, and accounts for its further develop-

ment into new knowledge. Activity (Tätigkeit) is shaped by (a) historical and cultural forms of production of material, intellectual, and spiritual collective life and (b) forms of human collaboration. Activity is not, as in many other theories, just a sequence of actions or a discursive turn-taking occurring during conversation or social interaction.

8 In a 1943 letter to Rudolph Kolisch about Beethoven's music, while he was in exile in Los Angeles, Adorno writes: "And like you, I believe in the strict knowability of music—because music is itself knowledge, and in its way very strict knowledge" (Adorno, 1998, p. 180).

CHAPTER 4

Learning

1 Outline

In the previous chapter I presented the concepts of knowledge and knowing as they are conceived in the theory of objectification. This chapter focuses on the concept of learning. This concept is central to sociocultural theories. However, as we shall see, its formulation is far from easy. In the first part of the chapter, I discuss the concept of learning as participation in social practices and Vygotsky's concept of internalisation. Then, I introduce the concepts of objectification and learning.

2 Learning as Participation in Social Practice

In sociocultural theories that use *participation* to explain learning, the fundamental idea is that individuals acquire cultural knowledge as a result of their participation in social practices. There is an explicit intention to go beyond the individualistic conceptions offered by mainstream psychology and philosophy in which the individual is the unit of analysis and the focus of research.

The idea of participation was developed by Rogoff (1990), Lave (1988), and Lave and Wenger (1991), among others. Rogoff, for example, conceives of learning as *apprenticeship* in a context of guided participation. Rogoff says, "The concept of guided participation attempts to keep the roles of the individual and the sociocultural context in focus" (Rogoff, 1990, p. 18). She then states that she uses the analogy of apprenticeship "to focus on how the development of skill involves active learners observing and participating in organized cultural activity with the guidance and challenge of other people" (Rogoff, 1990, p. 19).

In discussing learning as apprenticeship, Rogoff shows how children and parents engage in subtle processes of shared attention, and how with the help of adults, children gain *insights* into social referencing (e.g., the child's interpretation of adults' facial expressions), ways to solve problems, and ways to deal with social demands. She notes that "Through guided participation with others, children come to understand and participate in the skilled activities of their culture" (Rogoff, 1991, p. 191).

In the footsteps of Rogoff, and drawing on ethnographic research on craft apprenticeship among Vai and Gola tailors in Liberia, Lave and Wenger have

 | DOI: 10.1163/9789004459663_005

explored the concept of learning through the construct of legitimate peripheral participation.

> "Legitimate peripheral participation" provides a way to speak about the relations between newcomers and old-timers, and about activities, identities, artifacts, and communities of knowledge and practice. It concerns the process by which newcomers become part of a community of practice. (Lave & Wenger, 2009, p. 29)

Within this participationist approach, "the meaning of learning is configured through the process of becoming a full participant in a sociocultural practice" (Lave & Wenger, 2009, p. 29).

Looking closely, however, we realise that, in participationist approaches, learning remains in the end a process whose goal is to *adapt oneself to social practices*, to *insert* oneself into them. Despite the interesting array of new concepts that the participation accounts bring in, learning appears to be a kind of adaptation, much as in Piaget's account. The difference is that while for Piaget adaptation is carried out through universal cognitive mechanisms of knowledge formation that are the same "in all different periods of human history and in all children, irrespective of social group and ethnicity" (Piaget & Garcia, 1989, pp. 266–267), in the participation paradigm learning is the adaptation through social mechanisms (guided participation, peripheral participation, etc.) to a cultural world of practices.

Furthermore, learning is investigated as a general concept. This is a point that Lave and Wenger make clearly in their *Situated Learning* book. Their focus is, as they say, on learning *in general*. "We should emphasize, therefore, that legitimate peripheral participation is not itself an educational form, much less a pedagogical strategy or a teaching technique. It is an analytical viewpoint on learning, a way of understanding learning" (Lave & Wenger, p. 40).

In the theory of objectification, the interest is to explore learning not in general, but as it happens when students engage in learning mathematics. Although a general reflection on learning can shed light on the learning of mathematics, our hypothesis is that the investigation of learning is *affected* by the *specificities* of what is learned, namely *knowledge*—mathematical knowledge.

As it shall become apparent in the chapters of this book, the intention in the theory of objectification is to offer a sociocultural approach to the teaching and learning of mathematics that is not exactly of a participationist nature. In the next section I turn to Vygotsky's concept of internalisation that has often been associated with questions of learning in sociocultural approaches (see, e.g., Daniels, 2002; Kozulin, Gindis, Ageyev, & Miller, 2005; Vadeboncoeur, 2017).

3 Internalisation

The idea of internalisation or interiorisation was developed in the field of psychology by Pierre Janet (1929) and Vygotsky (1929) in the late 1920s. It is a theoretical construct to describe the relationship between the psychological development of the individual and her environment. Janet, for example, argued that all psychological laws have two aspects: an outer aspect (dealing with other people) and an inner aspect (dealing with oneself). Almost always, he says, "the latter is posterior to the former" (Janet, 1929, p. 288).

Internalisation (Вращивание – vraschivanie) is one of the central ideas of Vygotsky's historical-cultural theory. The idea is articulated in several of Vygotsky's writings including *The Problem of the Cultural Development of the Child* (Vygotsky, 1929), *Tool and Sign in the Development of the Child* (Vygotsky & Luria, 1994), and chapter 5 of *Thinking and Speech* (Vygotsky, 1987). Internalisation appears as a theoretical construct related to Vygotsky's conception of the development of the mind, a conception that the Russian psychologist expresses in terms of "genetic laws of cultural development."

The first law emphasises the distinction between (1) innate mental functions and (2) artificial mental functions; that is, those affected by the use of signs and artefacts. The artificial or higher mental functions are precisely functions mediated by material culture. Vygotsky states this law as follows:

> The first law of the development and structure of higher mental functions which are the basic nucleus of the personality being formed can be called the law of the transition from direct, innate, natural forms and methods of behavior to mediated, artificial mental functions that develop in the process of cultural development. (Vygotsky, 1998, pp. 167–68)

Vygotsky's second law asserts that collective social forms of behaviour become methods of adaptation and of individual thinking. Vygotsky writes:

> The second law can be formulated thus: considering the history of the development of higher mental functions that comprise the basic nucleus in the structure of the personality, we find that the relation between higher mental functions was at one time a concrete relation between

> people; collective social forms of behavior in the process of development become a method of individual adaptations and forms of behavior and thinking of the personality. (Vygotsky, 1998, p. 168)

Vygotsky reformulates this law as follows:

> Every function in the cultural development of the child appears on the stage twice, in two forms—at first as social, then as psychological; at first as a form of cooperation between people, as a group, an intermental category, then as a means of individual behavior, as an intramental category. This is the general law for the construction of all higher mental functions. (Vygotsky, 1998, p. 169)

Then, Vygotsky introduces the third law on the cultural development of higher mental functions: "The third law, connected with the second, may be formulated as the law of transition of a function from outside inward" (Vygotsky, 1998, p. 170).

The third law refers, therefore, to the transition from the social to the individual. It is here where Vygotsky introduces the concept of internalisation: "The internalization of cultural forms of behavior involves the reconstruction of psychological activity on the basis of sign operations" (Vygotsky, 1978, p. 57).

The concept of internalisation is one of the most known concepts of Vygotsky's work. It allowed psychology to rethink the role of society and signs and artefacts in human cognition. However, its transposition to educational problems of teaching and learning is not something simple to accomplish. There are several reasons. In what follows I mention four.

3.1 *The Problem of Focus and Content*

First, there is a problem of *focus* and *content*. Indeed, the *focus* and *content* of Vygotsky's concept of internalisation is not learning, but the higher psychological functions (such as memory, attention, and perception). The problem that internalisation seeks to explain is not how the child learns but how the higher psychological functions arise from social behaviour and social relations, how they are transferred to the inner or mental plane, and how they develop.

Of course, the concept of internalisation insists on the existence of a relationship between what is learned and the intellectual development of children. It might not be surprising hence, that, in the empirical investigation of internalisation, educational psychologists have turned to learning. Thus, in his article, *Stage-by-Stage Formation as a Method of Psychological Investigation*, Piotr

Iakovlevich Gal'perin (1992), one of the most prominent of Vygotsky's followers and a well-known scholar for his investigations of internalisation (Gal'perin, 1967), discusses three types of learning: trial and error, formal learning, and one based on the internalisation of a schema that orients actions and that reveals at the same time the attainment of a higher level of development. In this type,

> learning [is] directed toward acquiring a general method for studying objects, one that, in "mental form," represented a method of thinking of a kind the children had not had before, and that signified a transition to a qualitatively higher level of intellectual development. (Gal'perin, 1992, p. 77)

In this educational psychological line of inquiry, learning is seen as a problem of development—not as an educational problem *per se*. As a result, the focus remains to a large extent the question of cognitive development; the child is seen through cognitive lenses (i.e., as a "cognitive subject"), and (following the external-internal path of internalisation) the problem to be explained revolves around how sensible, material actions become "mental actions," or "pure thought" (Gal'perin, 1992, p. 63; see also Gal'perin's 1989 paper "Organization of Mental Activity and the Effectiveness of Learning").

3.2 *The External-Internal Planes*

Second, the concept of internalisation seems problematic in the sense that it involves a dichotomy between two planes: an internal and external one. This problematic stance can be better understood if we bear in mind the fact that Vygotsky's theory was developed as an attempt to overcome the reflexological and idealistic approaches of his time. Vygotsky often claimed that psychology inspired by reflexology investigations was a mindless behavioural psychology and that psychology inspired by subjective idealism (e.g., introspection) was a theory of mind without behaviour. On the basis of Spinoza's work (1989), Vygotsky tried to overcome dualist theories (theories based on two systems, the internal and the external) and to formulate a monistic theory of consciousness. However, despite his commendable efforts, the resulting theoretical elaboration of the link between the social and the individual was not free of contradictions. Veresov—considered one of Vygotsky's most important contemporary specialists—asks:

> What essentially does it mean to abandon the postulate of two system existence and to what conclusions and logical effects does it lead? This

> logically leads to a full rejection of the idea of the existence of the internal and the external and, consequently, to the radical refusal of the concept of internalisation as a mechanism of the origin of internal structures of consciousness. Actually, the concept of internalisation becomes senseless in this sense. (Veresov, 1999, p. 226)

This line of inquiry, based on the internal-external dichotomy, led Vygotskian theorists to try to find a way out of it.[1] Although these theorists agreed upon the idea that "intellectual activity ... emerges developmentally from the subject's external, practical activity" (El'konin, 1995, p. 32), they sought to "refute" (p. 33) the dualist opposition by showing the genetic unity of internal and external activities; that is to say, that they have the same structure, that the former is analogous to external, practical, activity. "We assumed," El'konin says, "that any internal activity was the orienting component of an external activity" (p. 33). However, El'konin argued, intellectual activity *is* an activity on its own. But then, this intellectual activity should have an orienting activity, as practical "external" activity does. We face here an infinite regress: the new orienting activity of intellectual activity will be an activity in itself, etc. What do we find at the end of the process? El'konin feared that psychology would be then reduced to logic. In the end, it seems that this line of inquiry, based on the premise of two planes—the internal and the external—leads to a sense that the solution to the problem of the relationship between intellectual and external activity was somehow unsatisfactory. As one of Vygotsky's students put it,

> Under Vygotsky's leadership we began to look at these questions, but then we somehow drifted away from them because other problems arose, and they were pushed into the background. Essentially, for a long time we were forced to be content with the fact that some external correlations were established between activity and mental processes. (Zaporozhets, 2002a, p. 49)

3.3 *The Individual as a Replica of Her Culture*

Third, the concept of internalisation seems also to be problematic insofar as the individual appears to be conceived as a *replica* of her culture. In fact, this is what follows from the genetic laws of cultural development. Indeed, as González Rey remarks, Vygotsky's notion of internalisation was associated with an instrumental-operational outlook of individuals and their world (Gonzales Rey, 2009, p. 62).

Naturally, Vygotsky did not have in mind a kind of mechanical transition from the social to the individual. Thus, in *Tool and Sign in the Development of the Child,* Vygotsky argues that

> The transfer of social methods of behavior to the interior of the system of individual forms of adapting is not at all a purely mechanical transfer; it is not done automatically but is connected with a change in structure and function of the whole operation and is a special stage in the development of higher forms of behavior. Complex forms of cooperation transferred into the sphere of individual behavior begin to function according to laws of the primitive whole, of which they now make up an organic part. (Vygotsky, 1999, p. 53)

Despite Vygotsky's claim concerning the non-mechanical nature of the external-internal transfer, the result was that "subjectivity and the subject [became] mere epiphenomena of discursive, semiotic, and linguistic practices" (González Rey, 2011a, p. 36). At the end of his life, in order to better understand the anchoring point between the individual and the social context, he came up with the concept of *perezhivanie*, "a concept which allows us to study the role and influence of environment on the psychological development of children in the analysis of the laws of development" (Vygotsky, 1994, p. 343). See also Veresov (2017). Yet, in the formulation of the laws of genetic development, there is not much room left for agency and subjectivity, making their theorising "one of today's major challenges in cultural-historical activity theory" (Stetsenko, 2020, p. 5).

3.4 *Internalising Relations between People*

Fourth, in addition to the above, the higher psychological functions are reduced to the internalisation of relationships—"concrete relation[s] between people," as Vygotsky says. Unfortunately, Vygotsky's concept of *relations between people* remains unspecified. And I think that, even if we try to specify it, we will not be able to advance very far. For example, how do we distinguish between the "concrete relations between people" (Law 2) that are at the origin of mathematical thought and those "relations" that are at the origin of legal thought? And even within mathematics, how can we distinguish between the relations between people that would lead to geometric thinking and those that would lead to probabilistic thinking? Something seems to be missing.

Certainly, in the sixth *of Theses on Feuerbach,* Marx says: "the essence of man is no abstraction inherent in each single individual. In its reality it is the ensemble of the social relations" (Marx, 1998, p. 570). This is how Marx overcame the substantialist ideas about the individual of the romantic post-Enlightenment trend of his time (for details, see Henry's (1976) remarkable book *Marx*). Roth (2018) argues that the adjective "social" qualifying the word "relationship" in the sixth thesis on Feuerbach should actually be translated as "societal." The difference is not purely terminological, but conceptual. The difference is important insofar as it involves in itself a key idea of the relationship between the individual and the social-cultural world as it is conceived by dialectical materialism. Roth says:

> "the societal" is taken as indexing something specific to society as a whole, whereas "the social" is taken as indexing any situation in which there are two or more people involved in some relation. Thus, whereas "the societal" implies "the social," the latter does not imply the former. (Roth, 2018, p. 3)

The social relations (or, rather, the societal relations) that appear hence in the sixth of *Theses on Feuerbach* mean that they are not merely contingent relations but cultural-historical through and through. They go beyond the relations that appear in each interaction in which the individual participates. Now, it is certainly clear that human relations in the sense of societal relations have to do in crucial ways with the manner in which each one of us comes to conceptualise the world. But this emphasis on the relational dimension does not seem to me to be sufficient to make such relationships *the* sole substratum of the higher psychological functions and even less the sole substratum of conceptualisation (mathematical, legal, or other).

Vygotsky's latest writings show his efforts to overcome these difficulties, in particular his search for a broad definition of the concept of meaning (see Chapter 2 above; see also Presmeg, Radford, Roth, & Kadunz, 2016). I will not discuss here Vygotsky's effort and what the concept of meaning contributes to the resolution of the problem he was trying to overcome, for my intention is only to show that Vygotsky's theory, based on the idea of internalisation, formulated in a period that contemporary Vygotskian scholars identify as Vygotsky's "instrumental period" or (as mentioned in Chapter 2) "instrumentalist turn" (see González Rey, 2011b) is not free from theoretical difficulties that have implications for our educational conceptions of learning.

In the following section I explore a different path than the one suggested by the concept of internalisation. My approach is not psychological. There is hence no attempt to solve the difficulties mentioned above. I leave it up to the Vygotskian psychologists to deal with them. My problem is educational and more specifically related to learning. I want to explore learning following another path suggested by Vygotsky: learning as a problem of consciousness (understood in a dialectical-materialist sense, as in *becoming conscious of, and transformed by, something*).

4 Processes of Objectification

If we conceive of knowledge as a *generative capacity* or *cultural potentiality*, as I suggested in Chapter 3, knowledge cannot be something that we can "possess" or that we can "own." Knowledge is rather something that exists in our culture and that we *encounter* in the course of our life. Our *encounter* with cultural-historical systems of thinking (mathematical, scientific, aesthetic, legal, etc.) is what I call *objectification*. Since such an encounter does not occur all of a sudden, objectification is rather a *process*. The following sections are intended to provide an account of objectification as a processual concept. The account is articulated around specific ideas about:

1. the *individual* involved or engaged in learning,
2. that which is to be learned (i.e., *knowledge*), and
3. the meaning of objectification.

4.1 *The Learning Subject*

As mentioned before, the idea of the *individual* involved or engaged in learning (the learning subject) adopted in the theory of objectification is different from the one featured in the student-centred pedagogy, which borrowed it from idealistic and rationalist philosophies. In such philosophies, the subject appears as a *given* entity: the subject is considered to be the source of thought, knowledge, and meaning. In his critique of Kant and the idealistic/rationalist stance, Hegel calls this conception the "original identity of the 'I' within thinking" (Hegel, 1991, p. 83). Section 47 of Hegel's *Encyclopaedia* begins precisely with a summary of this idealistic/rationalist position:

> In my consciousness I always find myself (α) as the *determining subject*, (β) as a *singular* or as something abstractly simple, (γ) as what is *One* and the *same* in everything manifold of which I am conscious—as *something-identical*, (δ) as something *that distinguishes me* as thinking from *everything outside me*. (Hegel, 1991, p. 90, original emphasis)

The idea that the subject is the *determining* subject means that it is an invariable part of experience. It is the "*One and the same* in everything manifold" mentioned by Hegel. This point is particularly relevant and important in educational discussions, as it leads to the following question: What is, in fact, the role of the social environment and the historical, political, economic, and cultural context in the making of the subject?

The rationalist philosophy of the Enlightenment tried to solve this problem by conceiving of the individual as a *rational being*—*homo rationalis*. Reason is considered *universal*—the same one anywhere. The anthropological differences we notice from culture to culture would be, according to this approach, no more than different expressions of the *same* rationality (Radford, 2020a). It is precisely the adoption of these ideas that enables Piaget (Piaget & Garcia, 1989) to state that the mechanisms of the construction of knowledge are invariable throughout cultures and time. With this assumption, Piaget (1970) can think of the development of the subject in logical, universal terms that take place in an orderly fashion at successive stages. Translated from philosophy to pedagogy, this concept of the subject leads to a rationalist concept of the learner. Canadian psychologist Jack Martin has brilliantly described the result of this transposition: the learner appears as an individualist subject who learns in a self-regulating way, unconnected to the social environment and culture. In Martin's words, the learner is

> an individual laboring in relative solitude ... When social factors are considered they mostly are framed as variables that mediate or influence what is predominately a highly individualistic pattern of development. There certainly is little here that might speak to the possible socio-cultural, political, and moral constitution of personhood. The central concern is for an individual actor capable of simultaneous action and reflection on this action, much like a stereotypic scientist in close scrutiny and judgement of experimental phenomena of interest. (Martin, 2004, pp. 193–194)

This individual, Martin points out, acts like a Cartesian thinker, a self

> isolated from its surrounds even as it is shaped and affected by circumstances and events, which while always influential are not seen as in any way constitutive of the core self ... [A self] that surveys the exterior landscape for signs of personal affirmation and possibilities for expression on the one hand, and clues to strategic action on the other. Moreover, its most vital resources are apparently available within its detached internality. It acts as a final arbiter over whether or not its strategies are effective ... This is a self that ... requires only a facilitative grooming to become more fully socialized and intellectually engaged. (Martin, 2004, p. 197)

According to the dialectic-materialistic perspective I adopt here, the subject is not considered as a *given entity*. Objectification is an encounter; but it is not a simple encounter of a ready-made subject with cultural-historical knowledge. The encounter is, above all, the *constitution* and *transformation* of the learning subject as a result of this encounter.

4.2 *Cultural-Historical Knowledge and the Learning Subject*

In addition to being articulated around a specific conception of the *learning subject*, the concept of objectification that I am articulating here rests also on a conception of *knowledge* that is different from the philosophical positions referred to above. I have mentioned in Chapter 2 that idealism can be defined as the identity between subject (S) and knowledge (K): K = S. In other words, knowledge is what the subject produces: its prolongation. There is no mediation between subject and knowledge. The subject constructs her own world and the ideas it contains. Enclosed in her own world, the subject can therefore think and conceive of only what she has herself produced. The subject is a synthesiser of representations that makes the "self" invariable across epistemic experiences. Since the synthesis of representations is and can only be purely subjective, knowledge resulting from this synthesis is not something that can refer to a real cultural world (an "objective" world *with respect* to a given culture). Commenting on Kant's epistemological position, in Section 46 of the *Encyclopaedia*, Hegel says that knowledge obtained through the unity of self-consciousness has nothing objective; the objectivity attributed to it is only *subjective* (Hegel, 1991, p. 89)—something supposedly shared or "taken-as-shared," as contemporary constructivists say (see, e.g., Cobb, Stephen, McClain, & Gravemeijer, 2001). In the theory of objectification, the situation is different: knowledge (for example, the different ways of thinking and doing things in a given culture) exists at the moment of our birth, independently of each one of us. Knowledge has been collectively produced by previous generations of individuals. From the viewpoint of the learning subject, it is not generated from "within." As Mikhailov points out, in the beginning, knowledge, skills, and abilities "exist without me" (1980, p. 200). They are *there*, in our culture. We encounter them in the course of our lives.

As we saw earlier, in Figure 6 (Chapter 3) the adult chimpanzee named Yo breaks coula nuts. With her right hand Yo places the nut over an anvil and, in a coordinated manner, holds the stone hammer with her left hand, while the young chimps to her left and right watch her attentively. The young chimps do not yet master the relatively sophisticated motor and conceptual skills that are required to accomplish the cracking of the nut. These skills *already exist in their culture*. They exist as a generative capacity of actions. However, for young chimpanzees, this generative capacity still exists only as *potentiality*, and will become part of the young chimps' repertoire of action and reflection after a long period of *learning* that includes intense practice and observation.

Now, for a chimpanzee born in a community where there is currently no nut-breaking practice, such generative capacity does not exist. This chimpanzee cannot *encounter* this generative capacity as young chimpanzees in Yo's community can. It would be necessary to wait for the process of cracking nuts to be created and recognised collectively; that is, the process of institutionalisation and cultural recognition of actions and manners of doing. It is as an outcome of a cultural process that these actions and manners of doing are established as *knowledge*. Biro, Sousa, and Matsuzawa (2006), for example, report the case of a chimpanzee community in Bossou, Guinea, where two adult chimpanzees break panda nuts; however, this set of actions is not taken up again by other chimpanzees in the community. Boesch, Marchesi, Marchesi, Fruth, and Joulian (1994) note that similar cases may have occurred in other communities: "nut cracking might have been invented at some time but has never become established" (p. 336). In our terms, the cultural process of legitimation of actions is not carried out, resulting in knowledge not being produced. In order for knowledge to be produced, a cultural process of legitimation must occur.[2]

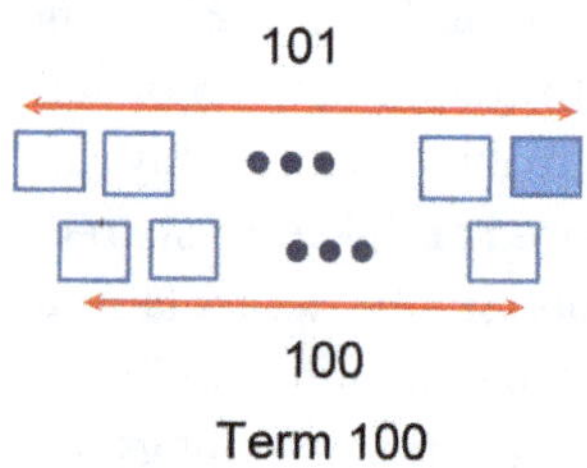

FIGURE 11
Frequently reported imagery of the term 100 for trained eyes

In the case of young students who are starting to learn specific math subjects (e.g., geometry or algebra), mathematical knowledge *already exists* in their culture. But these students do not necessarily master the relatively sophisticated motor and conceptual skills needed to investigate complex figures or extend arithmetic sequences. For example, mathematicians would easily identify the aspects of the terms in Figure 7 (see previous chapter) that are relevant to the generalisation task: they would probably see the terms as divided into two rows and would quickly notice the *functional* relationship between the term number and the number of squares in each of the rows. The perception of these variational relationships is usually so quick that mathematicians do not even realise the complex work behind it. Mathematicians would also easily extend the identified property to other terms that are not present in the perceptual field, such as Term 100; they would conclude that Term 100 has 100 + 101 squares; that is, 201 squares (see Figure 11). Even better, they would say that the number of squares of any term, for example n, is $2n + 1$.

The novice eye does not necessarily see the sequence that way. Figure 12 shows an example of how a Grade 2 student (7–8 years old) extends the sequence beyond the given four terms.

FIGURE 12 Terms 5 and 6 drawn by a Grade 2 student (7–8 years old)

The student focuses on the *numerosity* of the squares, leaving the *spatiality* of terms on the periphery. We cannot say that the student's answer in Figure 12 is wrong. The answer makes sense to the student, although it is probably true that, by focusing on the number of terms in the sequence, it may be more difficult to reach a general formula, such as 2n + 1. This is what we have observed again and again in our research with elementary school children and older students (13–17 years old). Older students tend to use trial and error methods that are not algebraic, but arithmetical (Radford, 2010).

The problem is not that students do not *see* the two rows of the terms. In Figure 13, we see a Grade 2 student facing for the first time a generalisation task of the type seen in Figure 7 (see previous chapter); in Figure 13 we see the student pointing with his pen to the top row, then the bottom row, after moving the pen between the rows to distinguish them. However, when the student draws Term 5, the *spatial* dimension of the terms is relegated to a second level; it does not play an organising role in the drawing of the term. The student draws *a row* containing all the rectangles of the term. The problem rests rather on not yet noticing that the spatiality of the terms provides interesting clues from an algebraic point of view.

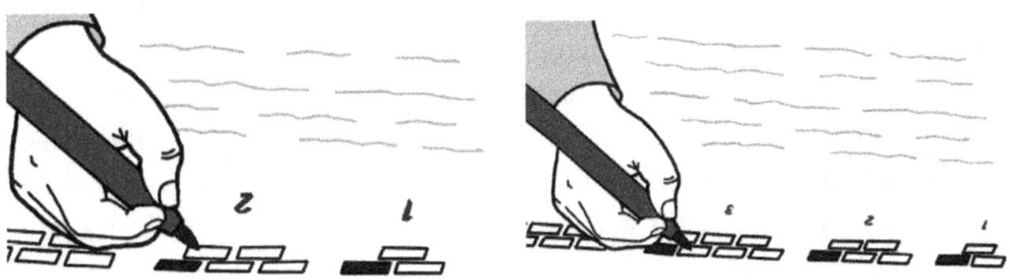

FIGURE 13 A student investigating the sequence. Thinking aloud, he points out the bottom row of Term 2 (left), then the top row of Term 3 (right)

The cultural forms of action, perception, and reflection that underpin algebraic thinking are not naturally invoked by the students. These forms exist potentially, but, for the students, they remain unrecognizable and unidentified. They are potentiality without actuality. To encounter and notice them, these forms of action and perception have to be put into motion. How? Through classroom activity. These forms will then appear in a sensible manner—as *knowing*. And it is in this sensible manner that, through processes of objectification, knowledge can become a theme of consciousness and thought, and learning may eventually occur.

5 Some Meanings of Objectification

It is worth mentioning two related but different German terms used by Hegel in his philosophy that have to do with the concept of objectification on which I am elaborating here.[3] The terms are *Gegenstand* and *Objekt*. Usually, both terms are translated as Object, although their meanings are not the same.

Objekt refers to an object in general, independent of the individual. On the contrary, *Gegenstand* means an object of consciousness, mediated by consciousness and continually changing with respect to it.[4] The difference is, then, that *Gegenstand* refers to an intentional object—an object of consciousness or thought—whereas *Objekt* refers to the object in its logical, non-subjective sense—an object (not necessarily material) in the world, a historical-cultural object (Inwood, 1992, p. 204).

Some translators of Hegel's work do make an explicit distinction between *Gegenstand* and *Objekt*. For instance, in his translation into French, Bernard Bourgeois uses "ob-jet" to translate *Gegenstand* and "objet" to translate *Objekt* (see Hegel, 1994). In their translation into English, Geraets, Suchting, and Harris do the same: they use *ob-ject* and *object*, respectively.

So, in Section 46 of the *Encyclopaedia*, when Hegel is criticising Kant for reducing the crucial concept of knowing or cognising as something that derives from abstract categories whose source belongs, in Kant's account, to the unity of individual's self-consciousness, he uses *Gegenstand*/ob-ject. The only object that the individual can know is a subjective object (hence an ob-ject with no fundamental relationship with a cultural object). In Geraets, Suchting, and Harris's translation:

> *To be cognizant*, however, means nothing else but the knowing of an ob-ject according to its *determinate* content. A determinate content,

> however, contains a manifold *connection* within itself and is the basis for connections with many other ob-jects. So, this [Kantian] reason has nothing but the *categories* for its determination of the *thing-in-itself*, or of that infinite; and when it wants to use them for this purpose, it *flies off* (and becomes "transcendent"). (Hegel, 1991, p. 89)

In Bourgeois' (1994, p. 89) translation, the first phrase is: "Or, connaître ne signifie rien d'autre que savoir un ob-jet selon son contenu *déterminé*."

In the original: "*Erkennen* heiβt nun nichts anderes, als einen Gegenstand nach seinem *bestimmten* Inhalte zu wissen" (Bourgeois, 1994, p. 88).

By contrast, when Hegel in the same section uses objective and objectivity as something beyond the will of the individual, he uses two terms related to Objekt: *Objektives* and *Objektivität*. In the following passage Hegel claims that what is known in Kant's account of knowledge and knowing does not bear any kind of objectivity, as it is the result of a synthesis of self-consciousness, and hence something purely subjective: "daß somit die Erkenntnis durch dieselbe[n] in der Tat nichts Objektives enthalte und die ihnen zugeschriebene Objektivität selbst nur etwas *Subjektives* sei" (Hegel, 1994, p. 90).

In Bourgeois' translation: "que par conséquent la connaissance obtenue grâce à elle[s] ne contient en réalité rien d'objectif, et que l'objectivité qui leur est attribuée est seulement quelque chose de *subjectif*" (Hegel, 1994, p. 91).

In Geraets, Suchting, and Harris's translation: "hence it is the view that in fact cognition through the categories contains nothing objective, and that the objectivity that is ascribed to them is itself only something *subjective*" (Hegel, 1991, p. 89).

We must not, therefore, confuse objectification in the sense of *Objektivierung*—which is part of the idea of *constructing* an object and making its reification—with the idea of objectification that I am articulating here. Nor should it be confused with the sense of objectification that appears in certain ontological approaches, in which the term refers to something objective that is universal and *independent* of human beings. The idea of objectification I propose is rather a *phenomenological* idea that has the meaning of *Vergegenständlichung*: a process that includes the individual's effort to apprehend an object (*Objekt*) that is already there (dynamic, changing object, always related to culture); a process in which the individual *expresses* herself precisely through the effort she makes to attend to the object (*Objekt*).

As the author of a dictionary about Hegel's philosophical terms tells us, in Hegel's Swabian dialect, *Gegenstand* means obstacle, which, being in front of me, resists me (Inwood, 1992, p. 203). Etymologically speaking, *Gegenstand* has its counterpart in the Latin word *ob-jacere*. That which *objects* to me: *objectare*.

It is from this sense of objectification, as something that stands before me and whose presence challenges me, that the theory of objectification borrows its name.

These conceptual distinctions will be helpful to define in the next section the concept of processes of objectification.

6 Processes of Objectification: A Processual Definition

Let us return to our definition of knowledge as *potentiality*, as a historical and cultural generative capacity for action and thought—for example, the mathematician's contemporary mathematical forms of action and reflection or the chimpanzee's forms of breaking nuts.

This generative capacity exists as something *latent* that we may (or may not) *encounter* in the course of our life depending on the cultural-historical-political webs of knowledge access ubiquitously operating in our society. *If* we encounter this generative capacity, it appears as something independent from us; it appears as an *Objekt*. Through intellectual and sensible activity, our consciousness may start noticing it and apprehending it from our subjective viewpoint. Using the terminology discussed above, we can say that objectification is a process in the course of which the cultural object (*Objekt*), hitherto unrecognised or encountered, is transformed into an object of consciousness (*Gegenstand*) (in the sense of Note 3, this chapter). The nature of such a transformation is that, by noticing it and apprehending it, the object and our consciousness *are transformed*. On the side of consciousness, new meanings appear. On the side of the object, knowledge (as *Objekt,* not as a *thing*, but as a *system of thought and action*) is transformed and becomes an object of consciousness. I shall come back to this point in a moment. For the time being, in order to specify further the concept of objectification, let me define it in more processual terms as follows: processes of objectification are the social processes of progressively becoming aware of cultural-historical systems of thinking and doing—something we gradually notice and at the same time endow with meaning. Objectification processes are those acts of meaningfully noticing something that reveals to our consciousness through our bodily, sensory, and artefactual semiotic activity. It is *noticing* or *perceiving* something in the course of our concrete practical activity through the emerging and evolving intentions and understandings as we project and cast themselves in the cultural-historical expressivity of semiotic systems, artefacts, and the kinesthetic movement of our body.

7 Learning as Objectification

The phenomenological meaning of objectification as *Vergegenständlichung*; that is, as a process in which the individual *expresses* herself precisely through the effort she makes to attend to, and make sense of, a cultural-historical system of thinking and action, allows us to envision learning as both processes of objectification and their outcome. Learning consists of noticing these cultural-historical systems of thinking and action. In metaphorical terms, learning consists of "touching" them or "grasping" them, as suggested by its etymology. In fact, learning comes from the Latin word *apprehendere*, composed of two prefixes: *ad*, which means towards, and *prae-*, which means *before*, and the verb *hendere*, which means *to catch* or *to grasp*. Learning is a movement forward to try to grasp something that lies before us. It is the complex encounter with knowledge and its subjective transformation into an *object of consciousness*.

The aforementioned transformation of an object (*Objekt*) into an object of consciousness (*Gegenstand*) that we place at the heart of objectification and learning does not entail an act of submission or subordination. We part ways here with participationist theories. Learning, in the theory of objectification, requires, in fact, a legitimate authenticity; that is, a *critical* look at knowledge and a space for dissent. *Learning, indeed, is the recognition of knowledge, but not its unconditional adoption*. That is why, from the critical perspective that I suggest here, teaching and learning are not seen as channels of social normalisation, but as processes that open up to subversive paths where new ideas and forms of action can be pondered.

8 Consciousness

Underlying the concept of learning introduced above is a particular concept of *consciousness*. Consciousness is not considered in the theory of objectification as a metaphysical construct hidden somewhere in an alleged interiority with which we would all have been born. This metaphor of "interiority" was invented towards the end of Antiquity. The Greeks of ancient times would have been scandalised by such an idea. For the Greeks, ideas (*eidos*) are not inside individuals, but in front of them (see Fried, 2009; Radford, 2004). The idea of interiority was developed by Augustine in a religious context and was later articulated by Descartes and his famous dichotomic view of the world: the one of interiority (mind, ideas, consciousness, etc.) and the one of exteriority (the concrete world) (for a detailed discussion, see Taylor, 1989).

In the theory of objectification, consciousness emerges from activity and it is in human historical-cultural activity that consciousness finds its substance. Leont'ev put it this way: the individual's "activity is the substance of his [sic] consciousness" (Leontyev (or Leont'ev), 2009a, p. 26). Activity gives form and content to consciousness so that, as Vygotsky (1979, p. 31) put it, "consciousness must be seen as a particular case of social experience." The result is that the structure of consciousness is the relation of the individual with the world (Vygotsky, 1997).

In this dialectical-materialist context, consciousness is, therefore, a subjective relationship and self-positioning with respect to the world. Consciousness is the emotional, affective, subjective process through which each one of us as individual subjects reflects on, and orients ourselves in, the world. This reflection is not a contemplative one. The individual consciousness is a specifically human form of subjective reflection of concrete reality in the course of which we come to form cultural sensibilities in order to ponder, reflect, understand, dissent, object, and feel about others, ourselves, and our world. Consciousness can only be understood as the product of cultural-historical and emergent contingent relations and mediations that, rather than being given, "arise in the course of the establishment and development of society" (Leont'ev, 1978, p. 79). Within this view, consciousness appears in concrete life, not as its origin, but as its result.

9 Teaching-Learning Activity

Learning/objectification is an *activity-bound* process. It occurs within the space created by the teaching-learning *activity* that transforms knowledge into knowing. And it is within the limits and possibilities of the manner in which, in this activity, knowledge appears as something sensible, that learning happens. As a result, learning will take one course or another depending on the teaching-learning activity from which it results.

Let us go back to the Piggy Bank example (Chapters 1 and 2). Let us think about this example taking place within a transmissive teaching-learning activity. The teacher shows the students how to calculate the money saved in Weeks 10, 15, 25. Then the teacher let the students answer similar questions. It can be argued that there is, within this activity, a process of objectification. There is, in fact, an encounter with cultural-historical knowledge. But the encounter is very limited. The division of labour that underlies the activity (the teacher says; the students listen and imitate) leads to a very restricted form of production of knowledge. Student involvement is very poor. There will not be much room to take a critical look at the ideas produced in the class (the teacher's ideas).

Let us now think of the same example in the context of a student-centred learning activity. In this case, the students are responsible for generating their own strategies. They get actively involved. They produce their own ideas. But there is not necessarily an encounter with cultural-historical knowledge. We saw in Chapter 2, for example, how students generated a non-algebraic strategy: the arithmetic doubling strategy. Until then, there had been no encounter with cultural algebraic knowledge, no consciousness of it and, therefore, no process of objectification. There is no encounter with culture and history; there is no encounter with the Other.

In the theory of objectification, the type of teaching-learning activity is different. It is what we have previously called *joint labour*. Joint labour implies a different conceptualisation of the division of labour in classroom activity and a different conceptualisation of the teacher and the students. In joint labour, learning consists of *actively* and *creatively* noticing and making sense of cultural-historical knowledge. In joint labour, teachers and students work together, but they do not do the same things.

If we look back at the Piggy Bank example, we see that in line 29 and onwards (Chapter 1) the teacher and the students start working together. The object of the activity is refracted differently in the teacher's consciousness and in the students' consciousness. The students, in fact, do not yet know where they are heading. The object of the activity is still unclear to them. But, in principle, this is no reason for not engaging in the classroom activity. Usually, the students' engagement in a teaching-learning activity has other motives that I discuss in another chapter. So, in the Piggy Bank example, Albert and his teammates engaged in the teaching-learning activity. In line 29 the students start noticing something new: the two columns of red bingo chips can be viewed differently. In a same term of the sequence, these columns have the same number of chips. Through language, the students name this act of perceptual consciousness, reaching a theoretical content that perception alone or action alone cannot offer. This is what happens when in line 30 Albert says, "It's always twice …" To emphasise the emerging awareness, the teacher repeats Albert's phrase (line 31). New conceptual possibilities appear:

30. Albert: It's always twice …
31. Mrs. Giroux: (*Repeating*) It's always twice.

Krysta reformulates what is progressively appearing to consciousness and tries to complete the sentence, since it remains to be defined as to what "twice" refers to:

32. Krysta: It's the double of what you ... No! (*She watches the artefacts intensely for a while and says*) I am confused!

However, the awareness involved in this process of objectification is not yet completely clear. Krysta admits to being confused. In the following line, Albert tries to formulate a more precise explanation, but the verbal formulation seems to be incomplete. Words are still missing. What Albert then produces is a *dynamic* embodied and material way of counting through which the functional relationship between the number of the figure and the number of red bingo chips begins to appear to consciousness:

33. Albert: Yeah! It's twice, look! (*Counting the red chips*) 1 + 1, 2; 2 + 2, 4; 3 + 3, 6; 4 + 4
34. Krysta: (*Interrupting*) 8.

The functional relationship between the variables remains *implied*. It has not been fully articulated yet.

Drawing on the concept of knowledge discussed in the previous chapter, we can say that there are certain *components* and *links* between the components of knowledge that begin to appear to the students' consciousness—such as the symmetry in the columns of red bingo chips; the fact that each column increases one unit from one term of the sequence to the next. That which has been noticed, those components and links of knowledge, should be noticed more deeply. To this end, the joint labour that embeds the process of objectification must continue. It will require an additional expenditure of material, intellectual, and emotional energy, and effort for the functional relationship between mathematical variables as *Objekt* to be transformed into an object of consciousness (*Gegenstand*).

10 Processes of Subjectification

As I mentioned above, in the encounter of the student with knowledge, a transformation of knowledge occurs, which is at the same time the constitution and transformation of the student's consciousness. This is why within the theory of objectification learning is not simply about *knowing*, but also about *becoming*.

In fact, in this perspective, learning is not *imitation*. Neither is it *participation* in a pre-established practice, nor the *insertion* of the subject into a foreign world, as often suggested by sociocultural currents that adopt the point of view of enculturation. Learning is not a process of inserting the world into the interiority of the subject, or the "transportation" of a process from an outer sphere to an inner one (Gal'perin, 1967, p. 29). In the theory of objectification, learning is the fusion between cultural modes of reflecting and doing and an emerging and fluid consciousness that seeks to perceive them (Radford, 2007, pp. 1790–1791). Learning is a continuous and tense encounter of mutual dialectical transformation between a cultural world (i.e., a cultural world that transcends the individual as a unique individual) and unique individuals who encounter it. In the course of this fusion, the world that appears to consciousness and the consciousness that arises from this encounter are continually transformed. It is for this reason that processes of objectification are entangled in *processes of subjectification*—processes of creation of a particular (and unique) self.[5]

11 Synthesis

In this chapter I introduced the concept of objectification. I emphasised the fact that the term objectification has several meanings.

a. One is the meaning of *Objektivierung*: objectification means to render something an object, as in transforming the actions of the subject into an object; that is, into something stable through a reification of actions.
b. Another sense of objectification is the one adopted by realistic or Platonist philosophies in which a transcendental reality is assumed as already given and *independent* of the individuals and *human activity*. Objectification appears here as something related to a universal truth, to something objective and bearing objectivity (in the sense of *Objektivität*).
c. The meaning adopted in the theory of objectification is different from the previous meanings. In the theory of objectification, the meaning of objectification comes from the Hegelian phenomenology: as object/entity that, in their ontogenetic development, stands in front of the individuals as an object that objects to them.

I started by presenting the phenomenological concept of objectification as an encounter with systems of thinking and their cultural meanings, systems and meanings that precede our appearance in the world. I then refined the concept as the process within which the transformation of cultural knowledge into subjective knowledge occurs, and observed that this transformation goes hand in hand with the creation and continuous growth of an individual's consciousness: objectification is a social process through which the individual progressively becomes critically aware of cultural-historical ways of thinking and doing; during this process consciousness is formed and trans-formed.

In this line of thought, objectification appears as an unending process, for, in their dialectical constitution, knowledge and consciousness are in perpetual transformative movement. We can always continue to understand, to try and imagine new things, to investigate things beyond where we are.

Learning, I suggested, consists of *actively* and *creatively* noticing these cultural-historical systems of thinking and action. I emphasised the adverbs "actively" and "creatively" as a reminder that the "noticing" in the definition of learning is not a contemplative act (as in the transmissive model), nor is it something purely mental (as it is in idealist accounts). It is something that requires effort and energy and that occurs simultaneously at the level of the individual and the level of the cultural-historical world. One of the forces of dialectical materialism (as developed by Marx and Ilyenkov after Hegel) is precisely to insist on the inseparability of culture, history, and the individual. Hence, *noticing* something in the course of our concrete practical activity, means that noticing is accomplished progressively and endlessly through our emerging and evolving intentions and understandings, as they unfold in activity and are expressed through our body and through language, graphs, and other semiotic systems according to their own cultural-historical expressivity.

But I also insisted that learning is a teaching-learning *activity-bound* process. From the viewpoint of the theory of objectification, it is not enough to be actively and creatively engaged in noticing. The teaching and learning activity should open a space for a critical positioning about cultural knowledge. There are hence not two but three adverbs qualifying learning: "actively," "creatively," and "critically."

What is it that distinguishes the concept of learning suggested in the theory of objectification from the ones found in transmissive pedagogy and in student-centred pedagogy? I mention here two elements that make a clear distinction.

The first one has to do with the concept of knowledge. In the theory of objectification, knowledge is not something that can be transferred, as if it were a thing or a commodity, from one person to another, as in transmissive teaching. Nor is it something subjectively produced, as in child-centred pedagogy. It is a cultural-historical entity always in motion.

The second element that distinguishes learning as conceived of in the theory of objectification from the conceptions of transmissive and child-centred learning is the use of the concept of consciousness. Learning is both becoming conscious and the creation of consciousness. Transmissive, constructivist, and other educational approaches do not need the concept of consciousness to define learning. If the concept of consciousness were removed from the theory of objectification, the theory would collapse.

This theoretical choice to define learning should not come as a surprise. Consciousness is a key concept in dialectical materialism. It runs across Vygotsky's work, from Vygotsky's paper presented at the Second All-Union Congress of Psychoneurologists, held in Leningrad in 1924 and published in 1925 (see Vygotsky, 1979) to the end of his life.

In the next chapter I will focus on some aspects of the practical and methodological investigation of processes of objectification and learning.

Acknowledgement

This is a revised and expanded version of Radford, L. (2013). Three key concepts of the theory of objectification: Knowledge, Knowing, and Learning. *Journal of Research in Mathematics Education*, 2(1), 7–44.

Notes

1 See also Arievitch (2003), Matusov, (1998), and Daniels (2002, pp. 39–44).
2 As can be seen, the process of *legitimation* makes knowledge something inherently cultural, social, political, and economic.
3 Hegel's henomenological terminology follows the philosophical tradition based on the key terms *subject* and *object*. Object, however, understood as object of knowledge or as object of consciousness, should be understood as a *thematic content*. So, to say that X is an object of consciousness means that X is a thematic content within the scope of human consciousness.
4 See the note of Hegel's translators Geraets, Suchting, and Harris in Hegel (1991, pp. xliii–xliv).
5 I discuss in more detail the processes of subjectification in Chapters 9 and 10.

CHAPTER 5

Processes of Objectification

1 Outline

The investigation of processes of objectification focuses on how cultural and historical forms of thinking and action become objects of consciousness in students. The conclusion we reached in the previous chapter was, that given the mediating role that activity plays in the transformation of knowledge into knowing, activity is a key component of the investigation of objectification processes. In this chapter we deepen our study of these processes, dwelling in more detail on the activity in which these processes occur. Some theoretical constructs related to the study of processes of objectification are presented. These constructs include: *semiotic means of objectification*, *semiotic nodes*, and *semiotic contraction*. The chapter ends with the introduction of the notion of *concept*.

2 The Investigation of Processes of Objectification

Let me start by recalling that the concept of activity has a specific meaning in the theory of objectification. This is what we call *joint labour* (the joint labour of teachers and students). I have already indicated that (teaching-learning) activity or joint labour includes material culture, such as mathematical signs and artefacts. Material culture is a bearer of historical intelligence, but it cannot by itself reveal the conceptuality that human work has deposited in them.

For example, in the Piggy Bank problem we discussed in Chapter 1, at the beginning of the lesson, we provided the Grade 4 students (9–10 years old) with bingo chips to address an arithmetic sequence problem. We gave them two coloured bingo chips (blue and red) and numbered transparent goblets representing the piggy bank in Weeks 1, 2, 3, etc., so that the students could model the savings process until Week 5. Then we asked them to generalise: they were required to answer questions about the amount of money saved in Weeks 10, 15, and 25.

The students began by modelling the savings process in the manner of a "real situation": they put the bingo chips *in* the goblets (three tokens in the goblet of Week 1, five tokens in the goblet of Week 2, etc.; see Figure 14).

 | DOI: 10.1163/9789004459663_006

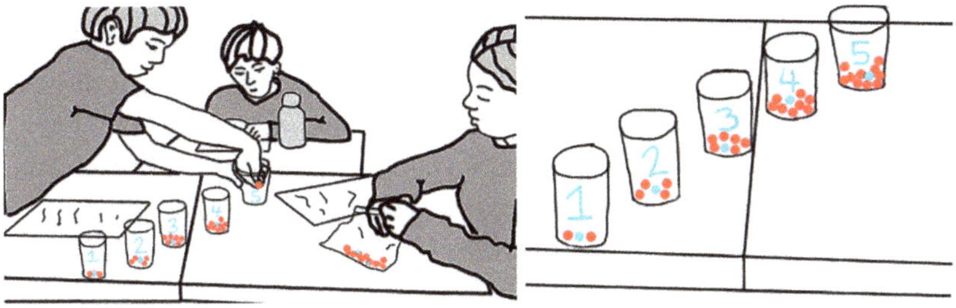

FIGURE 14 Modelisation of the learning process. The students put the tokens inside the goblets

Although the model was plausible, it was of little use in answering questions about the amount of money saved in distant weeks (such as Week 25). In fact, the students' attention was focused on *adding* two bingo chips each time. As a result, the bingo chips formed heaps in each goblet and were not linked to a more abstract multiplicative structure. The artefacts were insufficient to help the students reveal the general conceptuality targeted by the teacher's didactical project. The artefacts were carriers of a quotidian conceptual content far removed from the theoretical-algebraic content.

Thus, after a certain period of time, the students finished putting the bingo chips in the goblets without noticing any algebraic structure. The teacher was in the process of speaking to another group in another part of the classroom. I was filming one of the groups (Albert's group); I took off my headphones, left the video-camera on and went to talk to the group. I suggested that they put the bingo chips *in front* of the goblets. The students accepted my suggestion and began to stack them without distinguishing colours. Then I suggested that they use a blue chip to represent the initial dollar in the piggy bank. By following this suggestion, the three students of this group created a model of the savings process (see Figure 15; the goblet of Week 1 is not visible in the figure).

FIGURE 15 A model of an arithmetic sequence with bingo chips of two colours

The new layout of the concrete material helped the students to better understand the saving process. However, as we saw in Chapter 2, the students did not identify an algebraic formula to calculate savings in remote weeks.

The problem is that cultural-historical forms of thinking (in this example, algebraic forms of thinking related to numerical sequences) cannot be directly "seen." Algebraic knowledge *as such* cannot be shown in itself. It cannot be seen, touched, or felt. For knowledge to be something that can be perceived or sensed by human consciousness (that is, to be *intelligible*), individuals need to *do* something: it is, indeed, through an intellectual-material-sensible activity that knowledge can appear, and what appears is a materialised (or intelligible) developed form of it (this is the idea that the diagram in Figure 9 of Chapter 3 conveys). Of course, we could resort to the activity that underlies transmissive teaching, but, as I suggested in the previous chapter, this activity is very poor. It does not include the student, except as a passive observer. The kind of awareness that emerges from such an activity is therefore also very poor. This is not the kind of activity we resort to in the theory of objectification. The concept of activity that we want to nurture should endow algebraic knowledge with a particular *rich* conceptual content in a way that is *meaningful for the students*.

In this line of thinking it is important to distinguish between two "components" of teaching-learning activity. One component has to do with its didactical organisation (e.g., the choice of problems to discuss and the organisation of the problems). The second component has to do with the activity itself (activity as a moving *system*), something that, in the theory of objectification, encompasses much more than people interacting with each other. An activity is considered to be more than a *milieu* of interaction with people and artefacts. It is a *way of life*, something *organic* and *systemic*, an event created by a search with others for the solution to a problem, a search that is at the same time cognitive, sensible, material, emotional, and ethical.

In the next section these components of (teaching-learning) activity are introduced.

3 Teaching-Learning Activity

3.1 *The Φ Structure (or Didactic Component)*

The teaching-learning activity has an *object*. Prior to the activity, this object is identified by the teacher; it is part of her teaching project. This object can be the encounter of students with cultural-historical ways of thinking algebraically about sequences. It can also be the encounter with cultural-historical ways of thinking mathematically about movement, about time, fractions, etc. (see, e.g., LaCroix, 2014; Miranda, Radford, & Guzmán, 2007).

For the activity to develop in the direction of its object, one or more *goals* can be identified. If we continue with our algebraic example, these goals can be to solve problems about sequences in an algebraic way. To achieve the goals of the activity, a specific *task* can be devised. The task may consist of a sequence of related problems of increasing conceptual difficulty.[1]

The object-goal-task structure was suggested by Leont'ev (1978) in a more general context. Let us represent it with the letter Φ. I find Leont'ev's structure useful to provide a layered account of the pedagogical intention of the teaching-learning activity. However, the structure is not enough. In the theory of objectification, we add an epistemological analysis of the mathematical content that we find helpful to complement with an a *priori analysis* (Artigue, 1995, 2009).

In addition to the object of the activity, Leont'ev suggested the concept of motive, which is divided into the motive of the activity and the motive of the participants (Roth & Radford, 2011). The concept of motive is related to the concept of *need*; not as need itself but as something that moves us towards an objective.[2] "Motive," Leont'ev (1978, p. 171) says, "must be understood not as an experiencing of need itself but as that objective in which this need finds itself in given circumstances, which makes [that objective] an object and for that reason directs activity to a determined result."

In an attempt to reinterpret the structure of the activity, Davydov introduced another element: *desire*. Davydov starts by noticing that teaching and learning activity is a very particular case of activity: *learning activity*:

> A specific type of activity is investigated by Elkonin's school, that of learning activity. Activity being of many types, aspects and forms, it is difficult to orient oneself in its structure and content and work out an abstract theory unless one type of activity is extensively and thoroughly studied. (Davydov, 1999, p. 40)

He goes on to say that

> I will rely on the activity structure presented by A. N. Leontiev. He considers activity as consisting of needs, tasks, actions and operations. I agree with this activity structure but I extend it as follows: A. N. Leontiev did not consider a *desire* as an element of psychological structure of activity. In my view activity structure cannot be psychological, it is interdisciplinary,

> that is why only a certain aspect of it can be studied in psychology. (Davydov, 1999, pp. 40–41)

What does Davydov mean by desire? "A desire [*nuzhda*] is a core basis of a need" (1999, p. 41). The Russian psychologist then states that "nothing can be said about activity unless one understands spiritual or organic desire and how it is transformed into a need" (p. 41).

Although I sympathise with Davydov and find his idea of activity extremely interesting, I confess to not be convinced that, in an account of how learning happens, we need to rely on the transformation of desires into needs. I do think that we need to take into consideration both desires and needs in our investigation of learning, but I do not see why desires have to be transformed into needs. I think that desires and needs refract different aspects of the subjective life of the individuals. So, I take a different route—one in which the focus is not only learning (Davydov's *learning activity*) but also teaching. I am interested in *teaching-learning* activity.

3.2 *Teaching-Learning Activity*

In the previous section we have briefly discussed the *structure* Φ of the activity. Let us now look at the activity itself. The activity itself, understood as teaching-learning activity, is a *process*: one that *materialises* knowledge into something intelligible. Let us express it with the Greek letter Θ. Figure 16 presents the structure of the activity and the activity as a process.

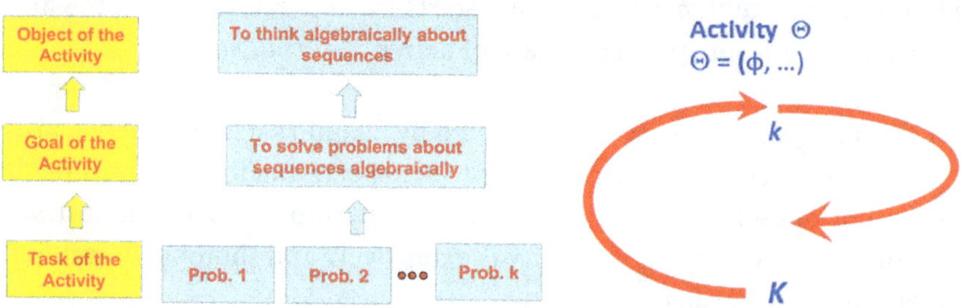

FIGURE 16 On the left, the structure Φ of the activity. On the right, the activity Θ itself; Θ is a function of the didactic project and other variables

The arrows on the left side of Figure 16 show how the task is aligned with the goal of the activity, which in turn is aligned with the object of the activity. In practice, usually, the object of the activity is defined, then the goal, and then the task with its various problems.

The arrow on the right side of Figure 16 represents the activity as a particular process: as a collective *form of life*, an *energy* that the participants release and that comes to envelop them, a spatial-temporal energy that is sensible and sensual, material and ideational, discursive and gestural, a flux carrying intentions, desires, and motive.[3]

Figure 16 attempts to capture the idea that the activity Θ that materialises knowledge (K) into knowing (k) is a process subject to change: it is what the discontinuity of the arrows suggests. As I mentioned in Chapter 3, while it is true that the activity mediates K and k, it is also true that K and k mediate the activity. This reciprocal mediation is precisely the meaning of the nature of the elements in a dialectical process.

The dialectical nature of activity can be better understood if we bear in mind the idea that an activity is a process situated in space and time which, although affected by the didactic project Φ, cannot be determined in advance. Teachers and researchers may have an idea, but the process is neither mechanical nor deterministic. The way in which activity occurs will depend on how students and teachers become involved in the activity, how they respond to each other, and their dynamic relationships to knowledge in general and to institutions, etc.

In the case of the theory of objectification, since the activity appears as a dynamic, complex system, we usually identify "moments" in the activity. In general, we divide the class into small groups of two, three, or four students. A first "moment" can be the presentation of the activity by the teacher (see Figure 17). Students then work in small groups. The teacher visits the groups and asks the students questions, gives them feedback, etc. (see "teacher-student discussion" in Figure 17). The teacher can invite two or more small groups to get together to discuss a particular idea (see "discussion between small groups"

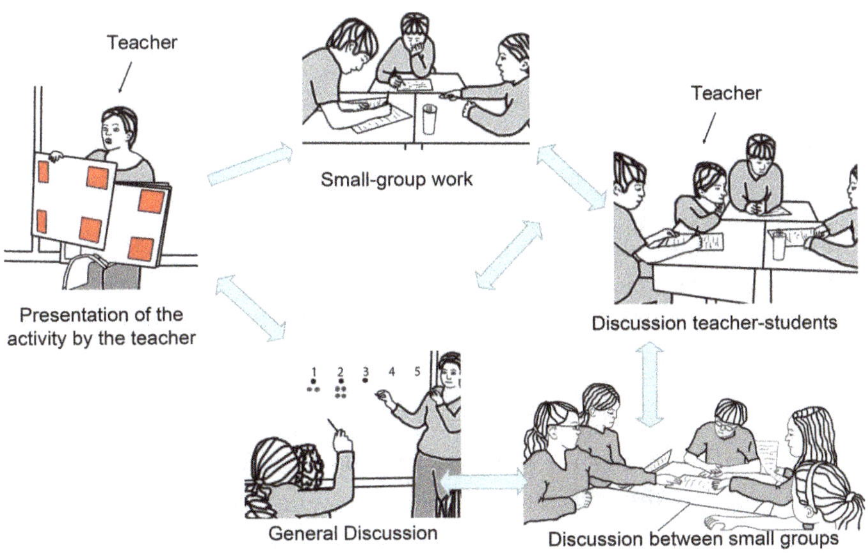

FIGURE 17 The activity Θ and some of its "moments"

in Figure 17). At some point, the teacher may invite the class to participate in a general discussion in which the groups can present their ideas and other groups can critically challenge them or make suggestions for improvement or generalisation (see "general discussion" in Figure 17). The class can end there or students can continue working in small groups.

A process of objectification occurs when students and teachers, through their joint labour, materialise the knowledge targeted by the didactical project—that is to say, transform it into something susceptible to be an object of consciousness—and the students start noticing or becoming conscious of it through such materialisation. In the Piggy Bank example, the process of objectification takes place when an algebraic way of looking at the savings sequence appears in the classroom. It is the process of transforming a cultural object (*Objekt*)—in this case, a cultural-historical algebraic form of thinking about numerical sequences—into an object of consciousness (*Gegenstand*).

As mentioned in Chapter 2, to distinguish the concept of activity that I am discussing here from other concepts of activity that exist in educational literature, we refer to activity as *joint labour*. Joint labour is not just a set of coordinated actions, but the activity carried out *side by side* by the teacher and the students. It is a form of energy that incorporates the flow of emotional, affective, ethical, intellectual, and material components and from which mathematics appears sensibly in the classroom. The concept of joint labour is a theoretical construct that allows one to move beyond the antagonism between teachers and students that has informed the transmissive teaching model and its behaviourist pedagogy as well as the various reform models and their child-centred pedagogy (Radford, 2014b, 2020b).

4 An Example of Investigation of Processes of Objectification

In our Piggy Bank example, after the students finished modelling the bingo chips as shown in Figure 15, they faced the question of the amount of savings in Week 10. As we saw in Chapter 1, at first, the students suggested doubling the savings from Week 5 and removing one of the blue bingo chips (see Figure 18, which is the reproduction of Figure 2 in Chapter 1). So, instead of the expected expression $10 \times 2 + 1$, the students suggested $11 + 10$.

After a discussion with the teacher, the students came up with a verbal formula for calculating the bingo chips in Week 5. The formula was "5 plus 5 plus 1." In Chapter 1, I presented the first part of this discussion between the teacher and the students. After that discussion, the teacher invited the students to

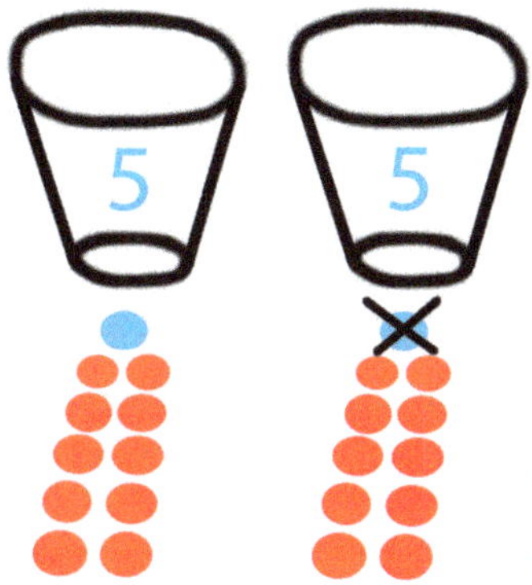

FIGURE 18
The students' strategy for calculating the savings in Week 10

write down the calculations to be made to determine the savings for Week 5 and Week 10. The teacher expected a formula like 5 + 5 + 1 and 10 + 10 + 1, respectively. However, the students produced nothing of the kind. They were entangled in attempts to combine the doubling strategy (see lines 22–28 in Chapter 1) and the emerging awareness of a relationship among the variables (see lines 29–40 in Chapter 1). The following excerpt gives an idea of the conversation with the teacher:

47. Mrs. Giroux: Which formula can we write?
48. Albert: (*Thinking hesitantly*) Hmmm …
49. Mrs. Giroux: What kind of calculations can you make?
50. Albert: Like times 2 or plus, times 2.
51. Mrs. Giroux: Week 5. (*She grabs the goblet of Week 5 and shows it to Albert*)
52. Albert: Times 2, no, it would be times 4. Five times 4.
53. Mrs. Giroux: Week 5.
54. Albert: Times 4 plus 1.
55. Mrs. Giroux: Why times 4?

The activity has led to an awareness of two ways of proceeding—one is Albert's approach, the other is the expected algebraic way that the teacher targets. These ways of proceeding lead to something that in dialectical materialism is called a *contradiction*. The contradiction is not considered a fault—a logical fault or a fault of the didactical design. On the contrary, contradiction is seen as the engine that pushes activity forward and keeps it moving. Contradiction is part of the movement of knowledge. Let us now look at the rest of the discussion.

After an exchange with the students, the teacher took the fifth goblet (Picture 1 in Figure 19) and asked Albert[4]:

67. Mrs. Giroux: What did you do here? 5 ... (*Pointing to the red bingo chips; see Picture 2 in Figure 19*) Times ...?
68. Albert: ... 2
69. Mrs. Giroux: (*Pointing to the blue bingo chip; see Picture 3*) Plus?
70. Albert: 1

Then Mrs. Giroux took the Week 5 goblet (Picture 4 in Figure 19), moved it to her left to a place where one would expect the Week 10 goblet to be if the sequence had been materially extended, and asked:

71. Mrs. Giroux: What would you do with Week 10, if there were Week 10 here? (*See Picture 4*)

Albert made an effort to respond but did not come up with the expected formula. As the videotaping analysis suggests, at this point, both the teacher and the student were very tense (see, e.g., Albert's body posture in Picture 5). Mrs. Giroux begins again:

77. Mrs. Giroux: (*Taking with her hand again the goblet of Week 5 and talking about Week 5*) What did you do here? (*Picture 6; the teacher is partially visible*)
78. Albert: (*Takes a long breath and hits the desk with the pen while Mrs. Giroux holds the goblet of Week 5; see Picture 7*) OK.
79. Mrs. Giroux: (*Still holding the goblet, pronounces softly*) 5 ...
80. Albert: (*In synchrony with Mrs. Giroux's gesture that points to the side of the red bingo chips; see Picture 8*) Times 2 ...
81. Krysta: (*Who is following the discussion*) Times 2 equal ...
82. Mrs. Giroux: (*Pointing now to the blue bingo chip; see Picture 9*) Plus 1.
83. Albert: (*Almost at the same time*) Plus 1.
84. Mrs. Giroux: (*Pointing now to an empty space where Week 10 would go; see Picture 10*) 10?
85. Albert: (*Mrs. Giroux silently points to where the red bingo chips should be; Picture 11*) Times 2.
86. Krysta: (*At the same time*) Times 2.
87. Mrs. Giroux: (*Silently points to where the blue bingo chip should be; see Picture 12*)
88. Krysta: Plus 1.
89. Albert: (*Looking at the teacher*) Minus 1? Times 2 minus 1? Plus 1?

Despite the inconclusive result of the interaction in lines 47–55 of the dialogue, in line 71 the teacher begins to try to engage Albert. She asks the

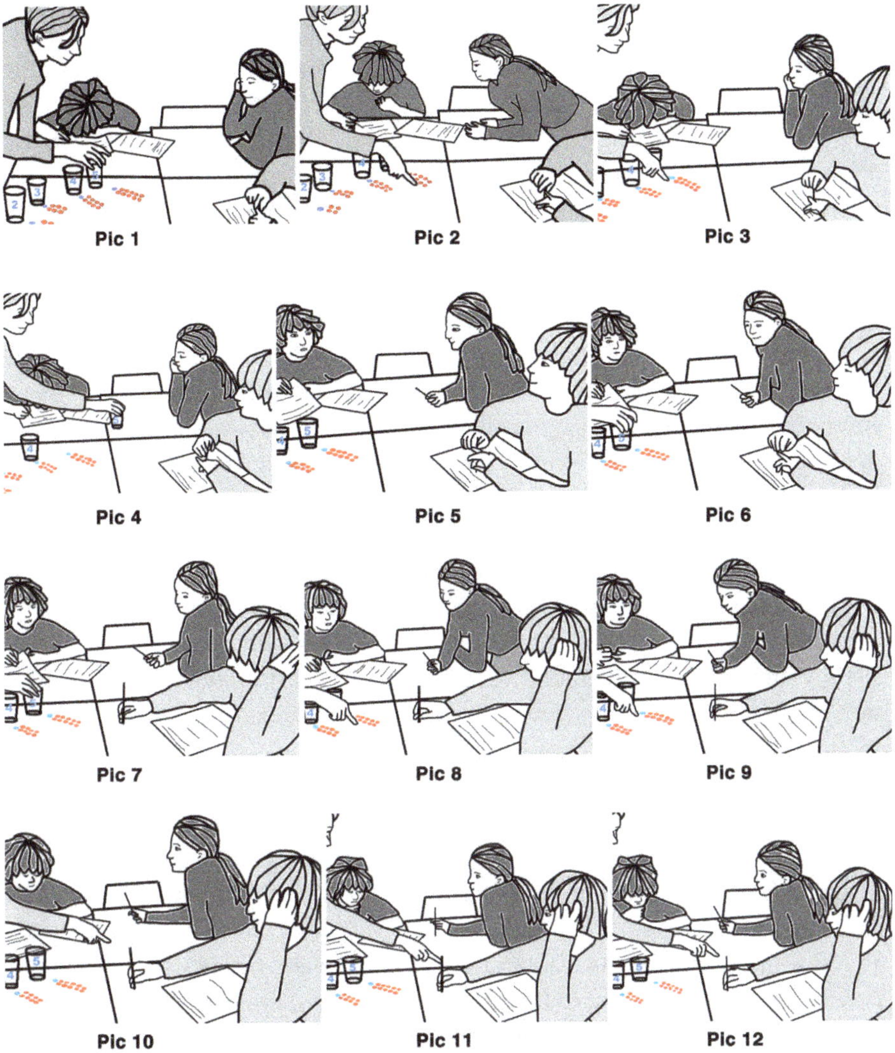

FIGURE 19 Mrs. Giroux and Albert engaged in joint labour

question, "What did you do here?" and then, in line 79, with an inviting word: "five," which she pronounces while holding the fifth goblet. Albert shows acceptance of the teacher's invitation with his whole body: he takes a deep breath and hits the desk with his pen (Picture 7). The way the teacher asks her questions encourages Albert: she implies that Albert can solve the problem but has not looked closely enough at the configuration of the bingo chips; he has not noticed what he is expected to see—that is, the mathematical structure from an algebraic point of view.

It is implicit that the teacher knows this algebraic structure. But it is not enough to know it. It is not enough because the teacher cannot inject this structure into the student's consciousness. For algebraic knowledge to appear, both the student and the teacher have to labour together. The teacher and the student must be involved in a process of objectification. This process will begin to take place when the sought-after knowledge leaves the realm of latent attention, ceases to be a potential, and its parts and connections begin to cross the frontier of explicit attention and begin to appear refracted in Albert's consciousness. But Albert and the teacher are not there yet.

Without speaking, in line 80, the teacher moves her hand to point to the red bingo chips (see Picture 8). Albert says, "Times 2." Then the teacher points to the blue bingo chip (Picture 9) and says, almost at the same time as Albert, "Plus 1." Then she moves her hand to the point where the model of Week 10 should be (Picture 10) and says softly, "10?" Then, without speaking, the teacher points to the imaginary position of the red bingo chips (Picture 11), while Albert looks at the indexical gesture and says, "Times 2" (see Picture 11).

This *energy* that envelops the teacher and Albert, energy that is sensible and sensual, material and ideational, discursive and gestural, and that is a flux carrying intentions and motives partially expressed and partially understood, *is* the activity or what we call joint labour. Wrapped in this energy, the teacher silently moves her hand to point with an indexical gesture the imaginary position of the blue bingo chip (see Picture 12) and Albert says, "Minus 1? Times 2 minus 1? Plus 1?"

At this point of the teaching-learning activity/energy/joint labour, Albert's encounter with aspects of algebraic knowledge is taking place. Albert still has to better perceive the different components and connections of knowledge; he has to better perceive the various elements of the mathematical formula. It won't take him long.

A few minutes later the teacher organised a general discussion. She invited several students to present their ideas. At one point she asked Albert to explain the calculations to determine the amount of money at the end of Week 2. The teacher was not sure that Albert had already become aware of all the elements of the formula. Maybe Albert was not sure either. The invitation was certainly risky. But Albert accepted the invitation and responded with some hesitation:

191. Albert: (*Referring to the second week*) Is 2; the second week is twice because they added 2 eh, dollars ...
192. Mrs. Giroux: Good ...
193. Albert: Then, 2 times 2 equal 4, which is the bingo chips, the red dollars.
194. Mrs. Giroux: OK. (*As Albert talks she identifies the corresponding drawings of the bingo chips on the blackboard; see Figure 20, Picture 1*).
195. Albert: 4 is twice the [number of the] week.

196. Mrs. Giroux: (*Repeating, to emphasise the idea*) Twice the week!
197. Albert. And 1, plus 1, like ...
198. Mrs. Giroux: (*To give a meaning to Albert's addition*) And you are going to add ... (*She points to the corresponding drawing of the bingo chip on the blackboard; see Picture 2*).
199. Albert: Yeah, plus one, which makes 5.
200. Mrs. Giroux: Good ... Do the same thing with Week 4. Same idea, 4.
201. Albert: 4 times 2 ...
202. Mrs. Giroux: 4 times 2 because it is twice ...
203. Albert: Plus 1, 4 times 2 plus 1 equal 9.

Throughout the general discussion, there is coordination between what Albert says and the teacher's gestures on the blackboard; the gestures are directed at Albert and the whole class, to give a visual sense to what Albert is saying (see Figure 20).

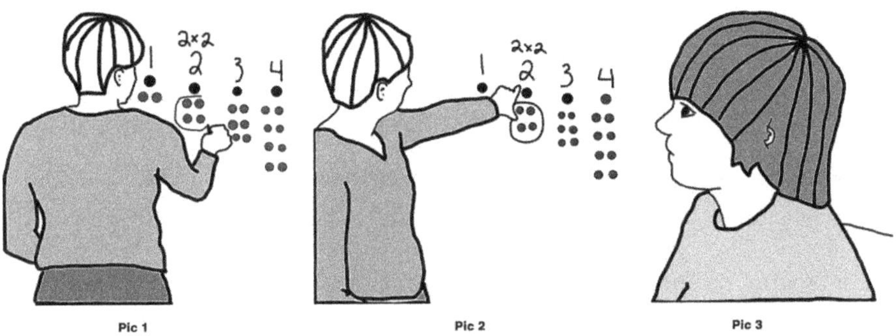

FIGURE 20 Coordination of gestures on the blackboard with what Albert says from his desk

In line 195, Albert verbalises the relationship between the variables. In line 196, the teacher takes the opportunity to emphasise Albert's idea. Line 197 shows the addition that was not clear to Albert in line 89. In line 198, the teacher helps to complete the meaning of the addition. Lines 200 to 203 contain the calculations for Week 4.

The lesson ended with this general discussion. The next day, the students worked on an isomorphic problem. The character in the story (the character that saves money) is not called Marc but Marianne. This time the piggy bank had $6 when Marianne received it and saved $3 a week, so by the end of the first week she had $9, by the end of the second week she had $12, and so on. While talking to his classmates about how to calculate savings at the end of Week 10, Albert said, "Add $3 each week. So we'll do it like this, 3 times 10 is 30 [plus 6] is 36. Yes, it's 36."

In a long process of objectification, which no doubt must continue in order to continue to refine the awareness of the algebraic way of thinking about sequences, Albert gradually managed to recognise the general mathematical structure underlying the saving process. Albert was able to encounter important features of a cultural-historical form of thinking algebraically related to the savings sequence, which was the *object* of those mathematics lessons. He was able to extend this awareness to new situations during an assessment that was conducted for the entire class more than a week after the end of the algebra classes. In the assessment the students were asked to find an expression for term 25 of the sequence shown in Figure 21.

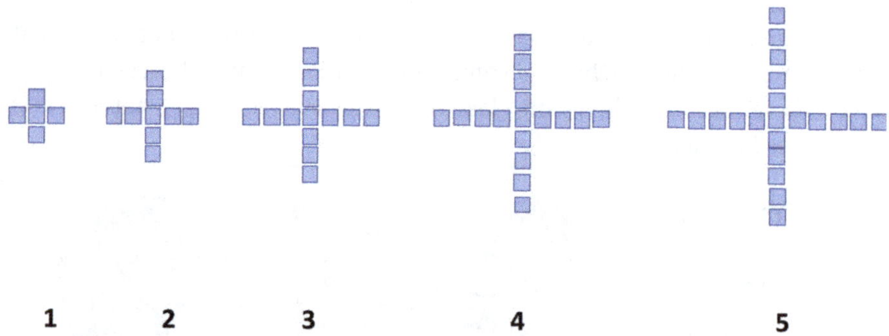

FIGURE 21 A sequence in an assessment that took place more than a week later

Albert's answer was: $25 \times 4 + 1$. He even did something else, as he suggested the following formula to calculate the number of squares in *any* term in the sequence:

$$\_\_ \times 4 + 1 = \_\_$$

In his own words, the first line in the formula "is for the number of the term." The number 4 means "the number of squares added each time. The number 1 is the first [square] we started with. The second line [to the right side of the equal sign] is the answer."

This example shows several concepts that I introduced previously.

4.1 *Joint Labour*

The example shows morphological characteristics of joint labour: we see how, as a dynamic system, joint labour appears as a spatial-temporal phenomenon whose texture is an energy that comes from the participants' efforts. This energy is sensible and sensual, material and ideal, discursive and gestural—a flow whose unpredictable course is driven by the dynamic of the affective, intellectual, and ethical dimensions that are configured as the activity occurs. It is worth repeating: this concept of activity as joint labour has nothing to do with the concept of activity of other approaches, where the activity is reduced, at best, to the coordination of actions of subjects and objects. In joint labour students and teachers engage with each other, work together and commit themselves responsibly.

4.2 *Knowing*

Knowing is what the teacher and students produce through their joint labour—what we called in Chapter 2 a *common work*. Phenomenologically speaking, knowing is this *intelligible entity* that, through joint labour, appears as a sensible phenomenon of thought and consciousness. In this example, *the intelligible entity* appearing in the classroom is the algebraic way of seeing, perceiving, thinking, and talking about the problem of saving. Knowing is not something psychological or mental. It is, like Beethoven's 7th Symphony as played and heard in a concert hall, a sensible entity. What appears—knowing—is a *singular*: something *always* different from execution to execution, from orchestra to orchestra, from conductor to conductor, from class to class. And, like music, what makes knowledge appear is the joint labour of the teacher and the students.

4.3 *Objectification*

The above example shows a process of objectification; that is, a process through which cultural knowledge (*Objekt*) is progressively transformed into an object of consciousness. This transformation, however, does not consist of bringing something from the outside into the inside. The movement of transformation goes in the opposite direction: the students *move towards something in front of them*. This transformation leads to a new way of perceiving, speaking, and conceptually addressing numerical sequences. This transformation is at the same time the transformation of consciousness, the emotional content of which appears clearly in the doubt expressed by Albert in line 89 of the previous dialogue and which is then replaced by a confident way of calculating. What results from a process of objectification is not the product of individual actions, nor the result of contemplation. The result of a process of objectification is *learning*. Learning is always a risky endeavour for it requires us to leave the comfort of our own solipsistic niche to turn to *Otherness*, to something other than ourselves, an unknown region where, however, we can make our home and feel at home (Russon, 2017).

In the rest of this chapter, I discuss four concepts that are important in the investigation of processes of objectification and the account of the students' learning.

5 Semiotic Means of Objectification

In processes of objectification, usually, teachers and students resort to a variety of artefacts, signs, and other devices. These devices are fundamental to the manner in which things become progressively noticed. These objects, tools, linguistic devices, and signs that individuals intentionally use in social meaning-making processes to achieve a stable form of awareness, to make apparent their intentions, and to carry out their actions to attain the goal of their activities, are called *semiotic means of objectification*. They are semiotic in that they are key pieces in the production of meanings embedded in the processes of objectification.

If we return to the Piggy Bank example, we see the important role that the goblets, the bingo chips, and the colour of the bingo chips played in the objectification process. They were important in producing a visual organisation of the saving model. In Picture 4 (Figure 19) we see how the teacher uses the goblet of Week 5 to help herself and the students to imagine the goblet of Week 10. The goblets, the bingo chips, and the colour of the bingo chips are examples of semiotic means of objectification. But Pictures 2 and 3 show also the role of the teacher's pointing gesture. In Picture 2 the pointing gesture points to the red bingo chips. In Picture 3, the pointing gesture points to the blue bingo chip. These pointing gestures are also examples of semiotic means of objectification.

There are other pointing gestures in this process of objectification. Indeed, Pictures 10, 11, and 12 show three pointing gestures that are different from those of Pictures 8 and 9. While in Pictures 8 and 9 the referent of the gesture is *visible*, in Pictures 10–12 this is not the case: the referent is not there. It is not in the realm of perception of the students and the teacher.

We can say that these pointing gestures fulfil a different indexical role. Following Bühler (2011), we can say that the first pointing gestures are *ocular deictics*: they show their referent, they perform a *demonstratio ad oculos*. The second pointing gestures, by contrast—those that point to nothing in the realm of perception—we can call *deictics am phantasma*.

Now, closer attention to the process of objectification under discussion shows also the presence of natural language, prosody (e.g., intonation), body position, and rhythm. All of them are semiotic means of objectification. They come to occupy different positions in the activity in general and in the process of objectification in particular. They do not work alone. They work together. They contribute to the unfolding global meaning and have different particular meanings as the process of objectification unfolds. Thus, *deictics am phantasma* in Picture 11 and Picture 12 refer to the red and blue bingo chips, respectively.

6 Semiotic Nodes

The systemic view of classroom activity adopted here leads to distinguish "moments" of the activity, which are defined in terms of what the teacher and the students are doing (e.g., working in small groups or having a general classroom discussion). In those moments the attention, awareness, and mathematical consciousness and understanding of the students are expected to be moving towards the object of the activity. But the movement of consciousness is not linear. In the path of the encounter with cultural knowledge, the students' consciousness go in different directions, sometimes noticing some of the components and connections of the object of knowledge, and sometimes in directions that are different from what is expected by the didactical object of the activity. What this means is that processes of objectification may or may not be occurring at a certain moment of the classroom activity. If we try to chart the processes of objectification in a given lesson, we may find that they occur at certain parts of the activity; they may be interrupted and started again later on. If we return to the episode discussed in Chapter 1, the students started generating a strategy that we may recognise as arithmetic, and not as algebraic (for a discussion of the distinction between arithmetic and algebraic thinking, see Radford, 2014c, 2018c). If we follow our definition of a process of objectification, there is no process of objectification in the part of the activity that goes from the beginning of the lesson up to line 28. There is a classroom activity but there is no process of objectification yet. The intervention of the teacher in line 29 changes the course of the students' attention, awareness, and consciousness. We can say that a process of objectification starts unfolding. If we follow the activity (which has become possible thanks to the capabilities of videotaping the students' classroom activity), we observe that after line 40 (see, e.g., lines 47–55 above) the students keep trying without success. When the students try to write down the calculations to carry out to find the number of bingo chips in Week 5, the relations between the variables is still not clear. In these lines there is no process of objectification. The process of objectification re-emerges in line 67. And it is different from the process of objectification in lines 29–40 (Chapter 1). In the terminology of Chapter 3, the components and connections of the object of knowledge have become clearer.

We can try to go further and pinpoint within a process of objectification those parts of the process in which the components and connections of knowledge are becoming clear to the students. In the example that we are discussing, the relationship between the variable "number of week" and "number of bingo chips" is becoming clear from line 77 to line 89. This segment of the process of objectification, in which semiotic resources of different kinds come to play a crucial role, is called a *semiotic node*. Another clear example of a semiotic node is the segment that goes from line 191 to line 203. What is interesting here is that the gestures are made by the teacher in coordination with Albert's language—with what Albert is saying (see Figure 20). A semiotic node is a kind of nucleus of a process of objectification.

In more precise terms, a semiotic node is a part of the joint labour where embodied and other signs from various semiotic systems are put to work together in processes of objectification. A semiotic node refers to *segments of activity* where students and teachers bring forward possible mathematical interpretations and courses of action against the backdrop of culturally and historically constituted forms of thinking and doing (Radford, Demers, Guzmán, & Cerulli, 2003; Radford & Sabena, 2015).

7 Semiotic Contraction

There is a phenomenon that I take as evidence of learning to which I would like to call attention. If we compare the segment of activity that goes from lines 191–199 to the segment of activity that goes from lines 200–203, we observe that things were accomplished faster. We observe the same phenomenon if we compare the segment that goes from lines 200–203 to Albert's utterance reported when he was dealing with Marianne's savings the next day (e.g., "3 times 10 is 30 [plus 6] is 36. Yes, it's 36").

We are witnessing here a *genetic* process in the course of which choices are being made about what counts as *relevant* and *irrelevant*. This process leads to a *contraction* of previous semiotic activity, resulting in a more refined linkage of semiotic resources. The significance of this process is that it reflects a deeper level of consciousness and intelligibility of the problem at hand. I take it as evidence of learning. The name of this process is *semiotic contraction*.[5]

8 Concept

The theoretical construct of *concept* is another key notion of the theory of objectification in its account of learning. Usually, a concept is thought of as a kind of mental entity. This is not the case in the theory of objectification where a concept is understood as an entity that is at the same time subjective and objective, material and ideal, and that is expressed in the practical and concrete activity of the student. A concept is something that is put in dialogue with the real possibility of action: something that enables the subject to do things in a cultural context with others.

A very long-standing Western tradition that started with Spinoza and that was continued by Kant and reappeared in Davydov's work is to see the concept of an object as a mental reproduction of the essence of the object. In Kant the idea appears as the reproduction of the object through rules or norms. This is his concept of the *schema* (Radford, 2005). Heidegger (1977), however, suggested that knowledge is not the simple application of concepts. In other words, there is no rule or norm capable of revealing in itself the essence of the object. For one thing, there is an aesthetic of the object that escapes the rule. For example, the concepts of "fair," "good," and "beautiful" cannot be reduced to rules or rules of application or production—something that lawyers and artists know very well. The weight of an object is a numerical measure of it. Regardless of its precision, it cannot reveal the measured object in its entirety. I adopt a theoretical position according to which the most intimate nature of knowledge is to be evasive, to be elusive. Regardless of their conceptual sophistication, concepts will always be precarious: they will always be unable to fully account for the knowledge they try to reveal. There is a surplus of knowledge, an excess that does not fully reveal itself in the concept. And this is also true of scientific concepts.

Thus, I think that, contrary to Spinoza (1989) and other philosophers and epistemologists who followed him, the object "circle" cannot be reduced to its rule or norm of production. There is a poetic dimension that escapes the rule. Behind my theoretical position is a very particular idea of the subject-object relationship. Instead of thinking of the object as something that gives itself fully to consciousness through a rule or norm, I think of this relation as a relation of *contact* in which the subject and the object come together and, in coming together, produce something *hybrid*. I think this is another way of understanding a central idea of dialectical materialism—and I believe that Vygotsky was following it when he seeks, in the *unity* of hydrogen and oxygen, not a compound, but something *new*. Something new arises from the dialectical relationship of two different things. Concept and individual merge dialectically and produce something *hybrid* (Moretti, Panossian, & Radford, 2018). This is the idea behind the notion of concept suggested here: a concept is something that enables us to engage in certain ways in the world, something that, like language, instead of being possessed, we come to enjoy. And by enjoying it, by tackling new situations, we change (for a more elaborate discussion of the concept of concept, see Radford, 2020c).

The data discussed in this chapter provide us with an example of concept. The relative facility and freedom with which Albert deals with Week 4 during the general discussion and with Marianne's problem the next day are possible because Albert has come to inhabit a concept, which allows him to deal with *other* similar problems as well—problems like the cross sequence that the students encounter in the assessment (Figure 21). A concept, in this sense, is something that applies to a range of situations. Albert inhabits a mathematical (algebraic) concept: the concept of algebraic generalisation of linear sequences, which includes the recognition of the involved variables, their functional relationship, etc. And reciprocally, the concept inhabits him. Although still fragile, this concept, originated in activity, becomes an "organ" of Albert's body. This is the hybridity I was referring to above. As a result of inhabiting a new concept, Albert has changed, as the data suggest. Indeed, the Albert of the first passages is not the same Albert of the last passages of the classroom activity. The concept connects Albert with culture and history and transforms him at the same time into a cultural-historical subject.

In this view of the concept and individuals, they need to be seen in motion, not statically. Concepts have borders, their borders move as the individuals who inhabit and enjoy them tackle and ponder new situations. Concepts change and by changing, the individuals who inhabit them change too.

9 Synthesis

In this chapter I discussed in some detail the concept of processes of objectification, which serves to define the concept of learning. Some ideas about how to investigate the processes of objectification were presented. Starting from ideas introduced in previous chapters, I argued that knowledge is made up of culturally and historically constituted forms of action and reflection. At first, in the course of our lives, these cultural-historical forms present themselves to us as simple potentialities—as generative capacities for action and thought. Such potentialities exist in culture and it is there that we find them for the first time. But we do not find them *directly*. In order to be found, these potentialities must be set in motion and made intelligible to consciousness and thought. In the case of geometry, for example, these potentialities include culturally and historically constituted forms of action and reflection about space and its objects. They include a conception of figures and a range of reflections on them (e.g., their cultural classification, how to represent them, how to measure them, how to establish truths about their properties, etc.). In the line of inquiry that I am following here, consciousness cannot grasp these potentialities *as such*. For instance, Albert (or any of us for that matter) cannot grasp a form of thinking as such. A form of thinking is a *whole* (a dynamic and always changing *whole*) and cannot reveal itself *in toto*. These potentialities may only appear in a *transformed* form. This transformation is the materialisation or the embodiment of knowledge, and it is what we have called *knowing*: the appearance of knowledge as a *singular* developed form.

What makes the transformation of knowledge into knowing possible is *activity*—human sensible, practical activity.

Confronted with an elementary algebraic, geometric, or other problem, the mathematician is capable of performing a sensible, material, and ideal activity with a view to solving the given problem. In the course of this activity, the generative cultural capacity is sensibly embodied through signs, diagrams, language, attention, theoretical perception, etc., that leave visible traces on which the mathematician can continue to reflect. Kant was both amused and curious that geometers start doing mathematics by rendering sensible what they are talking or thinking about. If the subject matter is about triangles, the geometrician, Kant remarked, "at once begins by constructing a triangle" (Kant, 2003, A716/B744, p. 579).

I mentioned earlier that when we ask mathematicians the number of squares in Term 100 of the sequence shown in Figure 7 (Chapter 3), the mathematician very often "sees" the terms of the sequence as consisting of two rows. Making a generalisation very quickly, she is able to imagine the next rows and to deduce that there will be 100 + 101 squares in Term 100. Knowing, that is to say, the sensible appearance of knowledge through this activity, occurs without problems. This is not the case of the student who is just beginning to encounter that cultural-historical form of action and thought that we call elementary algebra. The novice student still has to encounter these forms of thought, action, and reflection.

It is in this theoretical context that I suggested that learning is our *encounter* with these forms, an encounter that takes place in what we have called processes of objectification. These processes are defined as processes that occur within an activity through which, upon acquiring a *singular developed* form, knowledge as potentiality is shown. This phenomenological appearance of knowledge is, in effect, the only way for knowledge to become an object of consciousness for us humans.

The Piggy Bank example discussed illustrates these ideas. We introduced Grade 4 students to a series of increasingly difficult (saving) problems whose goal was the materialisation of a cultural-historical way of thinking that we recognise as algebraic. This way of thinking cannot just show up out of nowhere. It can only be materialised; that is, filled with theoretical content, by means of an activity Θ. Our didactic configuration of the activity—our component Φ (see Figure 16)—favoured a theoretical content in which a generalising formula was sought, using specific conceptual questions of increasing difficulty, material cultural objects (goblets, bingo chips, paper, pen), and sophisticated forms of social cooperation (see Figure 17).

The excerpts presented from the classroom activity show that the cultural-historical form of thinking initially went unnoticed by the students who resorted to arithmetic forms of generalisation. In order for them to notice the forms of algebraic thinking, the teaching-learning activity has to unfold in such a manner that the targeted algebraic ways of thinking become objects of consciousness; that is, that they can be *recognised*. Recognising here does not mean finding something we have already seen. To recognise is the establishment of a relationship of difference between the "I" and what is being recognised. More specifically, it is our attempt at overcoming the difference. According to Heidegger, "to *re*-cognize = to *differ*entiate, that is, something as that and that, and thus to grasp *it* as "*it*self" (Heidegger, 2004, p. 16, original emphasis).

"That" which Albert still seems to not perceive clearly is what appears shyly in lines 79–81 when Albert begins to notice that there may be a different way of looking at the bingo chips. A differentiation, that is a recognising, begins. The following intensive exchange between Mrs. Giroux and Albert, in which the two labour together, led to a materialisation of knowledge. In the midst of a joint process of objectification, the cultural-historical form of algebraic thought (*Objekt*) appears progressively to consciousness endowed with a specific theoretical content (it appears as *Gegenstand*).

Methodologically speaking, the identification of the semiotic nodes and the semiotic means of objectification mobilised by the teacher and the students provides a kind of window to the investigation of objectification processes. An example of a semiotic node is the segment from line 77 to line 89. Within this semiotic node we recognise the goblets, the bingo chips, the colour of the bingo chips, gestures, language, rhythm, and prosody as semiotic means of objectification. The investigation of the semiotic nodes and the semiotic means of objectification provides rich and important information about how learning is taking place.

The concept of semiotic contraction provides additional information on the processes of objectification that underlie learning. The semiotic contraction shows, indeed, how and to what extent the students are becoming conscious of the singular conceptual content of knowledge materialised in knowing.

Now, the awareness of this unique conceptual content leads to applying it not only to *this* or *that* piggy bank problem. Albert is able to apply it also to *other* problems, such as the problem addressed during the assessment, which has nothing to do with savings. In the terminology introduced in this chapter, Albert has come to inhabit a *concept*. A concept is the subjective version of cultural knowledge. It is the *refraction* of knowledge in the individual's consciousness through the sensible developed forms of knowledge that appeared in the activity. In our example, the refraction of knowledge in Albert's consciousness *transforms* him, and enables him to engage in new ways in the world. It is in this sense that the concept is a *hybrid* entity where organic and inorganic (cultural-historical conceptual) nature produce a new Albert, who, although different from the beginning of the algebra activities, we keep calling Albert.

Acknowledgement

This is a revised and expanded version of Radford, L. (2013). Three key concepts of the theory of objectification: Knowledge, knowing, and learning. *Journal of Research in Mathematics Education*, 2(1), 7–44.

Notes

1 In the theory of objectification, the sequence is organised around the *conceptual and contextual unity of the activity*. This concept will be discussed in Chapter 7. For the moment, my intention is to highlight the structure that derives from the distinction between object, goals, and task.

2 This concept of need was discussed in Chapter 2 in the course of our discussion about the anthropological conception of the human.

3 Chaiklin (2019) offers an interesting discussion about the concept of activity in Soviet psychology, showing, in particular, how this concept appeared first as the activity of an individual and moved later to the activity of a collective.

4 Note that the first goblet on the left does not appear in the picture and Albert is partially visible.

5 For other examples, see Radford (2008b, 2014c). We will find other examples in the next chapter as well. In fact, semiotic contractions are not difficult to find. They are not rare phenomena.

CHAPTER 6

Embodiment

1 Outline

In this chapter I present an overview of embodiment in the theory of objectification and discuss in particular the ensuing concept of sensuous cognition. I start by mentioning some of the problems that embodiment theories face in their account of the embodied nature of thinking. I contend that claims about the embodied nature of thinking must come to terms with the problem of the relationship between the body as a locus for the constitution of students' subjective mathematical meanings and the historical cultural system of mathematical meanings conveyed by the school. In the account of embodiment that I suggest, the body, the senses, and the objects of sensation are not considered as given entities, but entities mutually transformed by cultural-historical activity. In this line of thinking, sensuous cognition refers to the idea that human cognition is a culturally and historically constituted multimodal sentient form of creatively responding, acting, feeling, transforming, and making sense of the world. I illustrate these ideas through some Grade 2 classroom episodes.

2 Introduction

Embodiment has become one of the most prominent topics discussed in recent research in mathematics education and beyond. Embodiment comes with a promise: to move us beyond the idealist Western tradition that posits the mind as the producer and generator of the individual's ideas. For theorists of embodiment, this idealist assumption is inaccurate if not wrong. Something very important has been led out of the epistemological equation: the *body*. And some claim that it is not only the body but also something that encompasses the body—*matter*. The promise of embodiment certainly looks fascinating and thought-provoking. Embodiment promises to take us on a new path to explore things about human cognition that, because we were looking the other way, busy trying to find the origin of ideas in the internal ruminations and mechanisms of the mind, we failed to notice.

The new path, however, is populated by a series of theoretical problems that the embodiment paradigm needs to overcome if it wants to move beyond idealism and classical empiricism. Indeed, the body, with all its materiality,

 | DOI: 10.1163/9789004459663_007

could still be considered as the refuge of the transcendental "I" of idealism.[1] In this case, the body would merely be a more expanded version of the individual's mind, so that what was before considered "mental" appears now with a new name, "corporeal," but in the end we would still be talking about the same thing. What would have changed is our recourse to an enlarged vocabulary. Gestures, for instance, would be kinds of "windows" allowing us to get a glimpse into the individual's mind and how it produces its ideas. Or else, more in the vein of empiricism, the body could also be seen as that through which we produce knowledge. The genesis of knowledge would be found in what we touch, perceive, and smell, in the kinesthetic sensations of our sensuous experiencing of the world. But then, at the end of our journey, after having travelled the new path promised by the embodiment paradigm, we would find the ghosts of Locke, Hume, and other 17th and 18th century empiricists telling us that this is indeed what they were saying. They might acknowledge that our contemporary theorists of embodiment have pushed their ideas further, but, in the end, we would still be navigating in the waters of empiricism.

To really move beyond idealism and empiricism we need to come up with a cogent account of theoretical categories that are often invoked in theorising embodiment, such as what is meant by the conceptual and the embodied, the ideal and material, and their relationships, so that we can move beyond the enduring dichotomy between the senses and reason. The apparent absence of compelling accounts of these categories has led some critics, such as Terry Eagleton, to see in contemporary embodied perspectives of the mind no more than "the return in a more sophisticated register of the old organicism" promoted by 17th century empiricists and a token of "the post-modern cult of pleasure" and love for the concrete (Eagleton, 1998, pp. 157–158). What Eagleton fears is that contemporary embodied perspectives remain trapped in the sensing subject, and hence in a curious form of radical subjectivism with unclear possibilities to regain contact with culture and history (Eagleton, 1996).

Dwelling upon Vygotsky's (1987) and Leont'ev's (1974, 1978, 2009b) work, and Hegelian dialectical materialism (Ilyenkov, 1977), in this chapter I articulate the idea of sensuous cognition, a cultural-historical theoretical perspective where mind, body, and world are conceived of as intertwined entities. Sensuous cognition stresses the idea that our thinking, feelings, deeds, and in fact all our relations to the world (hearing, perceiving, smelling, sensing, etc.), are *historical intertwines* of our body and material and ideational culture. As a result, our senses change along with changes in our material and ideational culture (knowledge, spirituality, etc.).

In light of what has been said in previous chapters, the relevance of embodiment in the theory of objectification is easy to guess. The processes of objectification that underpin teaching-learning activity occur in sensible and critical *encounters* with knowledge. In these encounters, to bestow meaning on what is being encountered—i.e., knowledge—teachers and students engage in embodied ways. Embodiment appears hence in the theory of objectification as a form of *touch* arising in an encounter. But instead of assuming that what is to be touched is *just* there, and instead of assuming that the one who is touching is also *just there* too, so that the touching is a kind of contact between two already given entities, the touching we are referring to here is something that produces and transforms the toucher and the touched. The sensed object of touch produces the sensing subject, and, in a dialectical movement, the sensing subject produces the sensed object. The touching, the third element standing in relationship with both the subject and the object is, as can be intuited from the general conceptual form of the theory of objectification (see Figure 9, Chapter 3), *activity*—practical, concrete, material human activity. Embodiment in the theory of objectification is, then, an effort to overcome the primacy of the subject; it is also an effort to overcome the primacy of the object, the world.[2]

3 The Intertwining of the Senses and Culture

In reading Feuerbach's work, Marx noticed that his countryman philosopher was making a dubious assumption. Feuerbach was adopting a naive perspective on the materiality of the world. I have already written about this problem above (Chapter 2, Section 3.8), when I mentioned the case of the cherry tree, which, in the eyes of Feuerbach, appears in front of us in its full splendour, with sensuous certainty. Ludwig Feuerbach was a Hegelian converted to empiricism. In the *Philosophy of the Future*, Feuerbach argued that

> The real in its reality or taken as real is the real as an object of the senses; it is the sensuous. Truth, reality, and sensation are identical ... Only through the senses, and not through thought for itself, is an object given in a true sense. (Feuerbach, 1966, p. 51)

The *Theses on Feuerbach* is an ensemble of eleven propositions (or theses) that Marx wrote in a notebook. They are not a finished product; they are rather sketchy notes. Here we are interested in the first thesis. The first sentence of the first thesis reads as follows:

> The chief defect of all previous materialism—that of Feuerbach included—is that things [*Gegenstand*], reality, sensuousness are conceived only in the form of the *object*, or of *contemplation*. (Marx, 1998, p. 572)

In this sentence, Marx criticises Feuerbach by assuming that the object of sensation is merely given, as it could lend itself unproblematically to the sensing subject. What is hence missing here? The problem is that, in emphasising the object, the subject is not taken into consideration. Let us see the whole sentence:

> The chief defect of all previous materialism—that of Feuerbach included—is that things [*Gegenstand*], reality, sensuousness are conceived only in the form of the *object*, or of *contemplation* but not *as human sensuous activity, praxis*, not subjectively. (Marx, 1998, p. 572; translation slightly emended)

Feuerbach's materialism *separates* the subject and the object. Cultural objects, like the objects of production, for instance, appear as objects *as such*. This kind of materialism fails to grasp the distinction between *Objekt* and *Gegenstand* that I mentioned in Chapter 4; that is, the distinction between objects of culture and objects from a subjective viewpoint (such as objects of consumption or objects of aesthetic delight). This was Feuerbach's dubious assumption in Marx's eyes.[3]

The previous sentence is followed by the following sentence, where the critique is now addressed to idealism. Indeed, to fix the problem, idealism emphasised the subject as an active agent: "Hence it happened that the *active* side, in contradistinction to materialism, was set forth by idealism" (Marx, 1998, p. 572). However, Marx argues, the fixing move missed its target, because what idealism did was to consider the individual and its activation or activity in an *abstract* manner. Here is the full sentence:

> Hence it happened that the *active* side, in contradistinction to materialism, was set forth by idealism—but only abstractly, since, of course, idealism does not know real, sensuous activity as such. (Marx, 1998, p. 572)

The individual appears in an abstract manner, for the individual is considered as acting freely, according to her whims and her natural powers. It is in this way, indeed, that we find the individual pictured in many contemporary accounts of embodiment. Sheets-Johnstone's work is such an example. For her, we are endowed with a kind of primal animateness and original kinetic spontaneity

> that infuses our being and defines our aliveness, is our point of departure for living in the world and making sense of it. It is the epistemological foundation of our learning to move ourselves with respect to objects, and

> thus the foundation of a developing repertoire of "I cans" with respect to both the natural and artefactual array of objects that happen to surround us as individuals in our particular worlds. (Sheets-Johnstone, 2011, p. 117)

Thanks to this primal animateness, the individual moves and, in doing so, is converted into a corporeal "world-constituting consciousness" (Sheets-Johnstone, 2011, p. 128). The objects of consumption, to continue with our example, are posited as merely subjective, unhistorical, and independent of the cultural production that subsumes them.

As we can see, idealism makes the same mistake as Feuerbach's materialism: it also *separates* the subject and the object. While idealism puts the emphasis on the subject, Feuerbach's materialism puts the emphasis on the object. For Marx, things, reality, sensuousness as conceived by idealism, Feuerbach's materialism, and previous materialisms (e.g., Democritus's) end up in a mutilated materialism

> to which we must oppose a complete materialism, that is to say a materialism that restores the broken link between subject and object—a materialism which requires to bring forward the concept of ... *praxis*, which combines both aspects. Only this concept is indeed able to make the connection between an object, a determinate element of natural reality, and a subject actively engaged in the work of transformation of the world. (Macherey, 2008, p. 40)

Praxis or activity (*Tätigkeit,* in Marx's German terminology) is not something abstract taking place with abstract individuals (epistemic subjects, cognitive subjects, etc.). Activity in the sense of *Tätigkeit* takes place with flesh-and-blood individuals in their *practical* and *sensuous* cultural-historical reality. Instead of merely unfolding *in situ*, speculatively, as something occurring naturally, out of the blue, activity (*Tätigkeit*) is, on the contrary, a *historical* process. Activity, taken in this sense, follows indeed "a movement that is not only natural but historical" (Macherey, 2008, pp. 40–41). "Instead of conceiving of the world as an objective whole given in a natural manner," human activity accomplishes "the concrete fusion of the subjective and the objective" (p. 41).

To sum up, Marx's concept of activity or praxis is an explicit effort to overcome both: (1) speculative idealism that endows consciousness with false constitutive powers, and (2) the false concrete materialist sensualism of Feuerbach that conceives of the senses and the concrete in an abstract manner, as if the senses were strict natural entities of the order of nature and nature alone. In this way, the first of the *Theses on Feuerbach* achieves an inversion of perspective moving the conception of the senses and reality from heaven to earth. It does so by replacing the speculative stance of idealism and materialism with the idea that object, reality, and sensuousness are entities that can only be understood against the backdrop of practical human activity; that is to say, human historical labour. We move here from a view of the senses, the body, and the material world in which the senses and the body are conceived of as naturally given or merely intertwined to a view according to which they are *produced* historically and culturally.

Referring to materialism in general, Fischbach notes that, "The sensible material on which materialists have based themselves (both sensible things and human faculties, human sensibility) has been taken as a de facto datum and not as the sensible product of a human activity itself sensible" (Fischbach, 2015, p. 31). Explaining the meaning that human sensible faculties acquire in Marx's account, Fischbach continues:

> [Human] sensible faculties such as hearing or sight are not given identical to themselves from all eternity … they are produced, they result from and are modified by human activity, they participate in this activity and are an aspect of it. (Fischbach, 2015, p. 31)

The senses and their cultural, historical, and political development are intertwined with the objects they produce and attend to, not as transcendental objects or as merely objects of touch and perception, but as objects of human activity. By encountering these historical-cultural objects created by previous generations, humans create human sensibility and in doing so, humans express and confirm what Marx calls the human essential powers:

> Only through the objectively unfolded richness of man's [sic] essential being is the richness of subjective *human sensibility* (a musical ear, an eye for beauty of form—in short, *senses* capable of human gratifications, senses confirming themselves as essential powers of *man*) either cultivated or brought into being. For not only the five senses but also the so-called mental senses—the practical senses (will, love, etc.)—in a word, *human* sense—the humanness of the senses—comes to be by virtue of its object, by virtue of *humanised nature*. The *forming* of the five senses is a labor of humanised nature. The *forming* of the five senses is a labor of the entire history of the world down to the present. (Marx, 1988, pp. 108–109, original emphasis)

Cultural-historical activity dealing with nature, human objects, and reality, is hence the link between the subjective and the objective, between subject and object, and the ground where human senses are formed.

With these ideas in mind, I turn in the next section to the sense of perception.

4 Perception

Many years ago, in a landmark article, Lettvin, Maturana, McCulloch, and Pitts (1959) identified some image operations in the frog's visual system. Endowed with these basic operations (which include curvature and motion) the frog can, within certain limits, make perceptual distinctions. Thus, while it seems unable to see motionless prey and hence starves in a cage filled with dead insects (Roth, 1986), the frog is able to distinguish moving prey of different sizes (Anderson, 1993) and colours (Hatle & Salazar, 2001). With a considerably more complex visual system humans can discern a greater and more extraordinary panoply of differences and similarities. Were we humans left without this discerning capability, knowledge, knowing, and concept formation would simply be impossible. The world in front of us would be reduced to myriads of single and incommensurable facts: everything would be different from everything else and resemblances between things would be impossible to imagine. We would not be able to notice *resemblances* between *different* things and also *differences* between *resembling* things. But how do we learn to make distinctions? How do we learn to tell the similar from the different?

The basic operations of the frog's visual system are genetically built in and constitute what Lettvin et al. (1959) call, using a Kantian term, the *physiological synthetic a priori*—i.e., the physiological substratum that allows the frog to see the world the way it does. What about humans?

Research dealing with perception suggests that babies start noticing resemblances and differences in the early postnatal years. Although at birth the pupils are not yet fully dilated leading to a limited capacity for object fixation and discrimination, visual sensitivity develops gradually over the course of the first years. By 5 to 6 months, infants become visually aware of their environment; eye and hand movements become coordinated. By 5 to 7 years, the basic functions associated with the child's cortical sensory areas have completed their development. The child's basic sensory processing abilities match the adult's (Atkinson, 2000; Farroni & Menon, 2008). At this age, we can assume in principle that the child can see the world as the adult does.

In fact, this is not exactly the case. What could be called the human's *physiological synthetic a priori* turns out to be highly sensitive to the social context and the development of other sensorial organs that come to affect the way we see the world. The development of vision includes indeed several cortical and subcortical areas. As a result, like all sensory experience, visual experience "can influence the way the brain wires itself up after birth" (Farroni & Menon, 2008, p. 5). In the end, what we see is not the result of direct inputs but of stimuli already filtered by meanings and information about objects and events in the world—meanings conveyed by language and other cultural semiotic systems. Thus, in contrast to the frog's perception, rather than being a purely biological act, human perception is a social process through and through. Human perception is, as Wartofsky put it, "a cultural artifact shaped by our own historically changing practices" (1984, p. 865).

The so-called wild or feral children (children who, for some reason, have lived out of society) provide us with another example that illustrates the cultural-historical shaping of the senses. One of these children was found in the woods of Aveyron, between Montpellier and Toulouse in France, in 1800. The young child is described by Humphrey as follows:

> [he] trotted and grunted like the beasts of the fields ... was apparently incapable of attention or even of elementary perceptions ... and spent his time apathetically rocking himself backwards and forwards like the animals at the zoo. (Humphrey in Itard, 1962, p. vi)

The so-called wild boy of Aveyron—or Victor as he was called later—was placed under the care of Dr. Jean-Marc-Gaspard Itard, who designed a series of exercises to teach Victor how to speak and to catch up with the development of his intellectual faculties. Thus, the boy started distinguishing incrementally between the sound of a bell and that of a drum, went on to discern among the tones of a wind instrument, and later distinguished between vowels. Itard comments:

> It was not without difficulty and much delay that I succeeded at last in giving him a distinct idea of the vowels. The first that he distinguished clearly was O, next the vowel A. The other three presented greater difficulty and for a long time he confused them. (Itard, 1962, p. 58)

The training of the senses of sight and touch followed the training of the sense of hearing. While from a developmental viewpoint Victor was not able to catch up completely, the example shows clearly that without life in society the raw biological ensemble of orienting-adjusting reactions with which we are born remains undeveloped. As we live in society, interact with others, and participate in more or less specialised forms of training, the biological orienting-adjusting reactions undergo cultural transformation and are converted into complex historically constituted forms of sensing (e.g., ways of seeing, touching, hearing, and tasting), leading to specific features of human development and the concomitant forms of cultural reflection (Radford, 2014a). Reflection, as understood here, cannot be considered a passive act of receiving sensorial impressions, as

17th and 18th century empiricists hold. Reflection involves both: (1) something that transcends the organism as such (something that, in order to differentiate it from the subject itself, we can call "objective," namely the object (*Objekt*) of reflection), and (2) the reflected object (*Gegenstand*), something that is subjective (in the sense that reflection depends on the specific organism reflecting the environment). In this account, reflection is a relational category between subject and object that keeps them intertwined with each other. Reflection's phylogenetic evolution gives rise to a range of sensuous relationships between the organism and the world (thermal and other forms of sensation, movement, and so on). In the case of humans, the intertwining of subject and object, mind and world, acquires a specific dimension by virtue of their historically evolved capacity to interact with the world and among species' members.

The understanding of the manners by which we come to perceive concrete things and generalise them as we make the experience of the world is, of course, a major theme of educational research. That there are many such manners of perceiving, some definitely incompatible with others, is largely attested to by results from research conducted in different fields such as history, sociology, anthropology (Dzobo, 1980; Kawagley, 1990) and mathematics (e.g., Bowers & Lepi, 1975; Crump, 1990; Harris, 1991; Lancy, 1983; Lizcano, 2009; Owens, 2001; Pumuge, 1975). The issue is that despite what Piagetian and other rational epistemologies and their associated theories of human development have claimed, there are uncountable manners of abstracting and generalising the always individual and contingent facts intuitively given to us by the senses and filtered by culture. In the course of our ontogenetic development, the senses and our understandings *become* shaped in certain historically formed ways as we engage in sociocultural practices.

The classroom example that follows focuses on the ontogenetic transformation of the students' perception. It explores such a transformation as a result of participating in something that seems to be missing in frogs, namely symbolic and artefactual, collective, multilayered activity.

5 A Classroom Example

The following example comes from a longitudinal classroom study involving a Grade 2 class (7- to 8-year-old students). I shall focus on two passages from a generalising activity around a figural sequence that I have already mentioned in previous chapters. What I want to discuss is the manner in which the students' perception is transformed into a theoretical cultural form of perception required to tackle generalising questions.

5.1 *Intention and Perception*

At the beginning of a five-day activity, the students and the teacher explored some sequences together. Figure 22 shows one of the sequences used in the classroom mathematical activity.

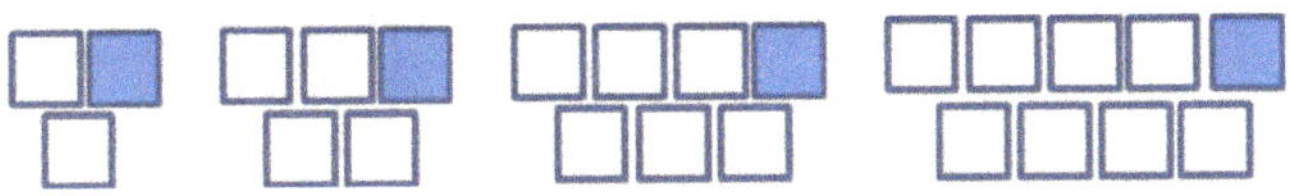

FIGURE 22 The first four terms of a sequence investigated in a Grade 2 class

In the first problem, the students were required to continue the sequence up to Term 6. In the next problem they had to discuss whether or not the drawing of Term 8 by Monique—an imaginary Grade 2 student—was exact (see Figure 23). The students were invited to discuss together and come up with an answer and a reason to justify their answer.

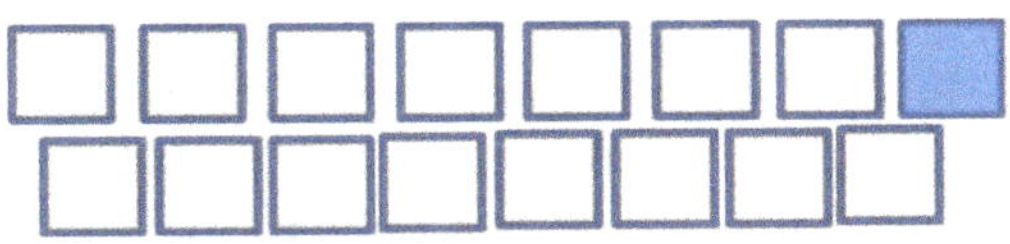

FIGURE 23 The term drawn by Monique. Is this Term 8 of the sequence?

As mentioned in Chapter 4, in dealing with problems 1 and 2, some students focused on the numerical relationship between consecutive terms, noticing that there were two more squares between one term and the next. They did not take advantage of the spatial clues suggested by the arrangements of the squares in each term. Figure 24 shows two examples.

As shown in the middle part of Figure 24, Terms 5 and 6 of the sequence were drawn as each having a *single* row. In other cases, the terms were drawn as having *more than* two rows (see Figure 24, right). Why? Contrary to what empiricist psychology has claimed, the image of an object in consciousness is not the simple mapping of the object. What is grasped of an object in a perceiving act is not the object in its totality. This is true even of "simple" objects, for they present to us many attributes (colour, shape, weight, odour, and so on). As Levinas (1989) remarked, the essential character of perception is to be *inadequate*. This is why it is not enough for the students to have the terms before their eyes. Instead of being complete, human perception is selective or, as Husserl said, *intentional*. In perception, Husserl (1931) argued,

> I am turned towards the object, to the paper, for instance ... Around and about the paper lie books, pencils, ink-well, and so forth ... but whilst I

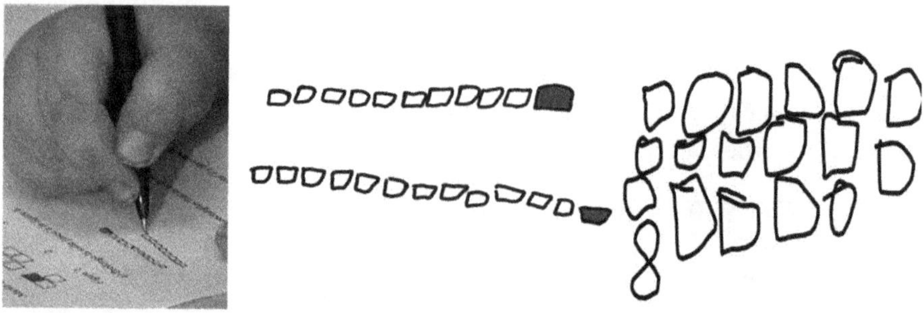

FIGURE 24 Left: the moment in which a student (James) is drawing Term 6. Middle: James's Terms 5 (top) and 6 (bottom). Right: Term 8 according to another student (Sandra)

> was turned towards the paper there was no turning in their direction, nor any apprehending of them, not even in a secondary sense. (p. 117)

To apprehend the books, an intentional act (a form of *intuiting* it) has to come to the fore. In the same way, to apprehend the terms as divided into two rows, a specific intentional act has to arise, for "'intuiting' [an object] already includes the state of being turned towards" it (Husserl, 1931, p. 117). The problem for the students, then, is to attend to the terms in a certain intentional way. They have to go beyond the intentional stance focused on numerosity, which makes the terms appear in a certain way in consciousness, to a different one, based on rows.

It seems that here we enter a vicious circle. To perceive a feature *O′* of an object *O*, a certain intentional act *I′* is required—an intentional act that will make available the intentional object under the form of its feature *O′*. The intentional object already requires a distinctive "*turn of thought*" (Husserl, 1931, p. 122, original emphasis). But the point is, and Husserl (1970) repeats it over and over, that the intentional act or manner of intuiting the object and the perceived object do not constitute two distinct aspects of perception: they are together the very basic unit of perception. To put it differently, intention and object *co-emerge* in the perceptual process of objectification.

The question, however, remains: how do the students move from a "mundane" or everyday apprehension of the terms to a different perception, for instance, a theoretical one? The question can be answered in two different ways. Since it is not just a matter of looking harder to the terms for the theoretical structure to become disclosed, one can argue that the mathematical problem or situation, if well designed, should by itself promote in the student the kind of scientific intention required. In other words, the intention and the mathematical structure should derive from the student's engagement *with the mathematical problem, with the mathematics*. Adopting a strong rationalist epistemology, this way of reasoning supposes that the student's mind is somehow equipped with the necessary scientific intuitions to build intentions in a scientific way. The second answer rests on the idea of development: the students are not yet mature enough to perceive the scientific structure.

From the cultural-historical theoretical perspective that I am advocating here, none of these answers is satisfactory. The rationalist stance, upon which the first answer rests, makes it unconvincing. I mentioned above that recent research in anthropology and history casts serious doubts on the idea of a single "natural" line of conceptual development and questions the idea that human reason is moved by the logic of the Enlightened scientific-based Reason portrayed by Kant and the 17th-century rationalists that preceded him (Radford, 2008c). The second answer, which is the one adopted by Piagetian pedagogues, subjugates instruction to development. Vygotsky (1987) articulated a substantial critique of such a view and argued that the relationship in fact goes the other way around: development follows instruction.

5.2 *The Domestication of the Eye*

Yet, to the experts' eyes, perceiving the terms as divided into two rows may seem a trivial endeavour. And, as argued in a previous chapter, surely it is. But it is so only to the extent that the mathematicians' eyes have been culturally educated to organise the perception of things in particular *rational* ways. The mathematicians' eyes have undergone a lengthy process of domestication. It appears now as part of a cultural knowledge that they have come to inhabit. That the mathematician's perception is not the result of a "natural" process of evolution is proven not only by results from cross-cultural psychology (Geurts, 2002; Segall, Campbell, & Herskovits, 1966) but also by our young students' responses—which is confirmed by our current research in preschool settings. The domestication of the eye is a lengthy process in the course of which we come to see and recognise things according to "efficient" cultural means. It is the process that converts the eye (and other human senses) into a sophisticated intellectual organ—a "theoretician" as Marx put it (1988, p. 107). Of course, I am not saying that the students did not see two rows (as I argued in Chapter 4). They surely did. But they did not deem it important to recognise the terms as being divided into two rows. Geometric clues were relegated to the background of attention to yield space to numerical matters. The capacity to perceive certain things in certain ways, the capacity to intuit and attend to them in certain manners rather than others, belongs to those *sensibilities* that students develop as they engage in processes of objectification.

Let me now get back to the students. The students worked in small groups of two or three. When the teacher came to see for the first time the work of James, Sandra, and Carla, the students had worked together for about 31:50 minutes. They had finished drawing Terms 5 and 6, answered the question about Monique's Term 8 (which they considered to be Term 8 of the sequence), and tried (unsuccessfully) to find the number of squares in Term 12. Noticing that the students were dealing with the sequences by adding two rectangles each time, the teacher engaged in collaborative actions to create the conditions of possibility for the students to perceive a general structure behind the sequence:

1. Teacher: We will just look at the squares that are on the bottom (*While saying this, the teacher makes three consecutive sliding gestures, each one going from the bottom row of Term 1 to the bottom row of Term 4; Picture 1 in Figure 25 shows the beginning of the first sliding gesture*). Only the ones on the bottom. Not the ones that are on the top. In Term 1 (*She points with her two index fingers to the bottom row of Term 1*), how many [squares] are there?
2. Students: 1!
3. Teacher: (*Pointing with an indexical gesture to the bottom row of Term 2*) Term 2?
4. Students: 2! (*James points to the bottom row of Term 2; see Picture 2*).
5. Teacher: (*Pointing with an indexical gesture to the bottom row of Term 3*) Term 3?
6. Students: 3!
7. Teacher: (*Pointing with an indexical gesture to the bottom row of Term 4*) Term 4?
8. Students: 4!

FIGURE 25 The teacher suggesting a perceptual manner to see the terms of the sequence

9. Teacher: (*Taking a short pause and breaking the rhythmic count of the previous terms, as if starting a new theme in the counting process, she moves her hand far away from Term 4 and points with an indexical gesture to the hypothetical place where one would expect to find Term 8; see Figure 26*) How many squares would Term 8 have on the bottom?
10. Sandra: (*Hesitantly, after a relatively long pause*) 4?

FIGURE 26
The teacher pointing to the empty space where one would expect to find Term 8

In line 1, the teacher makes three sliding gestures to emphasise the fact that they will count the bottom row of the four given terms. The gestural dimension of the teacher–students' joint labour reminds us of the tactile experience of the students reported by Zaporozhets (2002b) in an experiment focused on the ontogenetic process of the cultural transformation of the senses. In that study, 3- to 5-year-old preschoolers were learning to discriminate between variants of two geometric figures: triangles and quadrilaterals. The students followed the contour of geometric shapes with their fingers. At first, the preschoolers were making a substantial number of errors. Then, they were invited to trace systematically with a finger the outline of the figure, paying attention to directional changes of the motions at angles, and accompanying the tactile exploration with side counting (1, 2, 3 …). The investigator reports that at this stage perception was accomplished through the tactile experience while the eye performed an auxiliary role. "Later," Zaporozhets says, "the eye developed the ability to solve these types of perceptual tasks independently, consecutively tracing the outline of a figure, as it was earlier done by a touching hand" (2002b, p. 31). During this process, the eye undergoes a transformative change: "initially, the eye motions have an extremely extensive nature, consecutively tracing the entire outline of the perceived figure and simulating its specifics in all details" (p. 32). In a subsequent stage, the eye's motions "gradually begin to decrease and to focus on the individual, most informative attributes of the object" (p. 32). In terms of the concepts discussed in Chapter 5, there appeared a *semiotic contraction*.

In our example, the tactile dimension is carried out instead with gestures through which the teacher suggests a cultural form of perceiving the terms of the sequence—one in which the mathematical ideas of variables and the relationship between variables are emphasised. In the previous passage, we encounter again the two kinds of indexical gestures that we discussed in the previous chapter.

Indeed, first, there are the indexical gestures that the teacher and the students make to point to something in the realm of vision and perception. This is the case of the gestures that point to Terms 1, 2, 3, and 4 (e.g., the gestures in Pictures 1 and 2 in Figure 25). These are the gestures termed *ocular deictics* that perform a *demonstratio ad oculos*: they show the visible pointed object. Second, we can also distinguish the indexical gestures that the teacher makes to point to something that is *not* in the realm of vision and perception. This is the case of the gestures that point to Term 8 (see Figure 26). Since the object that is pointed to is absent, in the previous chapter we called these gestures *deictics am phantasma*. *Deictics am phantasma* are crucial in the constitution of a new psychic reflection: *imagination*. I shall return to this idea during our discussion of the next episode.

As we see in the previous excerpt, the teacher does not gesture silently. Gestures are coordinated with utterances. This is why it might be more useful to consider the teacher's utterance as a *multimodal* utterance; that is to say, as a bodily expression that resorts to various sensorial channels and different semiotic registers (Arzarello, 2006).

In this case, the teacher and the students proceeded to a coordination of eye, different types of gestures, the terms of the sequence, and speech through a series of organised actions that orient the students' perception and the emergent understanding of the target mathematical ideas. In the terminology introduced in Chapter 5, these semiotic resources *are semiotic means of objectification*. They become organised in what we have termed *semiotic nodes*; that is, segments of classroom activity where a complex coordination of various sensorial and semiotic registers occurs in a process of objectification.

The investigation of the semiotic means of objectification and their corresponding semiotic nodes in classroom activity is a crucial point in understanding the students' learning processes. The concept of the semiotic node rests indeed on the idea that the understanding of multimodal action does not consist of making an inventory of signs and sensorial channels at work in a certain context. From a methodological viewpoint, the problem is to understand how the diverse sensorial channels and semiotic signs (linguistic, written symbols, artefacts, diagrams, etc.) are related, coordinated, and subsumed into a new thinking or psychic unity (Radford, 2012a). Such a methodological problem makes sense only against the background of a conception of the mind that overcomes the dualistic view of internal–external processes. In our case, the methodological problem makes sense against the background of a concept of the human mind as sensuous through and through (Radford, 2014a).

Yet, as line 10 intimates, the passage from Term 4 to Term 8 was not successful. The noticing of the algebraic manner in which sequences can be algebraically perceived is not yet occurring. The teacher hence decided to restart the process, with some important modifications, as we shall see.

As mentioned previously, Term 8 of the sequence was not materially drawn. In the previous excerpt, the teacher *pretends* that Term 8 is on the empty space of the sheet, somewhere to the right of Term 4. She points to the empty space, as she pointed to the other terms, to help the students imagine the term under consideration. During the second attempt, the teacher does not go from Term 4 to Term 8; this time she goes term after term until Term 8:

11. Teacher: We will do it again ...
12. Teacher: (*Pointing to Term 1 with a two-finger indexical gesture*) Term 1, has how many?
13. Carla: (*Pointing with her pen to the bottom row*) 1, (*Without talking, the teacher points to Term 2 with a two-finger indexical gesture; Carla points with her pen to the bottom row of Term 2*) 2, (*Again, without talking, the teacher points to Term 3 with a two-finger indexical gesture* (*see Picture 1 in Figure 27*)*; Carla points with her pen to the bottom row of Term 3*) 3, (*Same as above*) 4, (*Now moving to the hypothetical place where Term 5 is expected to be, she makes two gestures: the first points to the imagined Term 5* (*See Figure 27, Picture 2*)*; the second to the imagined bottom row of Term 5; see Figure 27, Picture 3*) 5.
14. Teacher: Now it's Term 8! (*The teacher comes back to Term 1. She points again with a two-finger indexical gesture to the bottom row of Term 1*) Term 1, has how many [squares] on the bottom?
15. Students: 1.
16. Teacher: (*Pointing with a two-finger indexical gesture to the bottom row of Term 2*) Term 2?
17. Students: 2!
18. Teacher: (*Pointing with a two-finger indexical gesture to the bottom row of Term 3*) Term 3?
19. Students: 3!

20. Teacher: (*Pointing with a two-finger indexical gesture to the bottom row of Term 4*) Term 4?
21. Students: 4!
22. Teacher: (*Pointing with a two-finger indexical gesture to the place where the bottom row of Term 5 would be*) Term 5?
23. Students: 5!
24. Teacher: (*Pointing as above*) Term 6?
25. Students: 6!
26. Teacher: (*Pointing as above*) Term 7?
27. Students: 7!
28. Teacher: (*Pointing as above*) Term 8?
29. Students: 8!
30. Sandra: There would be 8 on the bottom!
31. Teacher: Excellent! Let's see if she [Monique] has 8 [squares] on the bottom [of her term].
32. Sandra: (*Counting the squares on Monique's term*) 1, 2, 3, 4, 5, 6, 7, 8! Yes, she has 8!
33. Teacher: Very well. Now we are going to check the top (*Twice she makes a slow gesture to indicate the top rows of the terms*). We'll look at the top.

In the second episode, the teacher refines her gestures. *Deictics am phantasma* have become more prominent (see Figure 27, Pictures 2 and 3). The teacher and the students counted together the squares on the bottom row of Monique's term and realised that the number was indeed 8. At this point the functional relationship between variables started becoming apparent to the students. The relationship started being noticed. The teacher then moved to a joint process of counting the squares on the top row.

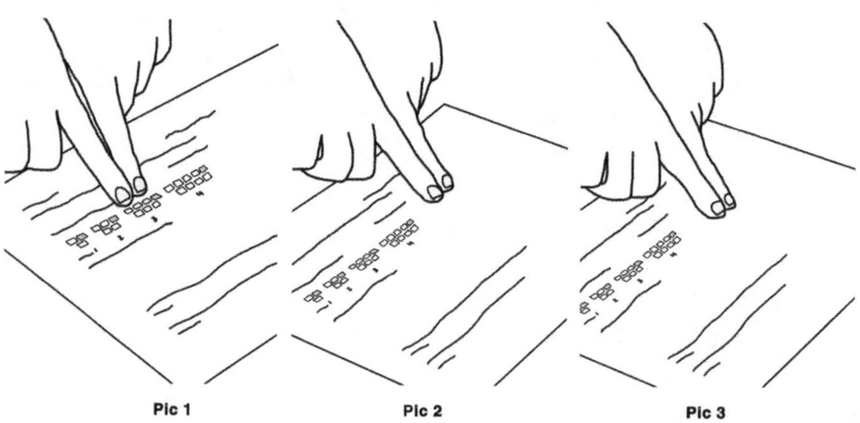

FIGURE 27 The teacher and the students explore the bottom row of the terms

The teacher repeated the same set of rhythmic pointing gestures as the students answered each of her questions. The counting was accomplished quickly (another instance of semiotic contraction). When they reached the top of Term 8 and figured out that there were 9 squares, the teacher invited the students to verify Monique's drawing. (The reader might have noticed that there are 8 squares in Monique's term; see Figure 23.)

The teacher pointed one after the other to the squares in the top row of Monique's term (see Figure 28a) while Sandra counted in a rhythmic way and the other students followed closely. Synchronising her counting with the teacher's gestures, Sandra said: "1, 2, 3, 4, 5, 6, 7, 8 ...!??"

The students were perplexed to see that, contrary to what they believed, Monique's Term 8 did not fit into the sequence. Here, the situation became tense. Figure 28b shows Sandra's surprise. Sandra and the teacher remained silent for 2.5 seconds; that is to say, for a lapse of time that was 21 times longer than the average elapsed time between uttered words that proceeded the moment of surprise.[4]

FIGURE 28 (a) The teacher points to the last square in the top row of Monique's term while Sandra says "8." (b) Astonished by the result, Sandra stays still and silently looks at the teacher for 2.5 seconds

6 The Poetry of Objectification

The previous interaction occurred in a process of objectification. This process consisted of the various perspectives brought forward by the students and the teacher and was shaped by the manner in which the participants progressively switched, revised, and refined their theoretical stances towards the mathematical problem at hand. Processes of objectification are indeed both relational and dynamic. They shift as the focus of attention shifts (Salinas & Miranda, 2018).

Now, while teachers cannot inject the object of knowledge into the students' consciousness, what they can do is create the conditions of possibility for the students to transform the object of knowledge into an object of consciousness.

In the classroom episode that I am discussing here, the teacher created the conditions of possibility for the students to perceive a general structure behind the sequence. And, as the example shows, the teacher did so by mobilising key semiotic resources. The shift of attention from strict numerical matters to a more encompassing view of the sequence based on geometric-spatial clues was made possible through an intense recourse to pointing gestures, words, and rhythm. While gestures allowed the teacher to point to the bottom row of the terms and words to qualify the number of the term, rhythm served the crucial cognitive function of creating a spatial-temporal order and expectation (You, 1994) that proved important in the synthesis on which generalisation rested. The disclosing of the mathematical structure and the concomitant poetic moment of objectification resulted from the complex link of those semiotic means of objectification (gestures, words, rhythm) that accompanied and oriented the students' perceptual, aural, linguistic, and imaginative activity.

The classroom setting permitting, there is, I want to argue, an aesthetic experience in the poetic moment of the encounter with the emerging mathematical structure. The aesthetic experience and the poetic moment have unfortunately been overlooked by rationalist approaches that reduce them to a mere "cognitive conflict." There is much more to thinking than the cogitative ruminations of the mind. Sandra was deeply surprised to see that what had previously seemed so obvious (i.e., that Monique's term was *certainly* Term 8 of the sequence) turned out not to be so. This aesthetic experience is the opening towards a new form of seeing. The teacher's sustained look into Sandra's eyes in the long silent pause that followed Sandra's discovery is a crucial form of communication in which two consciousness meet in front of the cultural mathematical meaning. Becoming aware of these poetic moments may empower us teachers and mathematics educators to offer the students room to enjoy these precious instances.

7 Counting the Unseen

The previous episode took place towards the end of the math class. The next day, the Grade 2 teacher started the math lesson with a general discussion. She drew the terms of the sequence on the blackboard and discussed with the class a counting method that, although similar to the one used in Sandra's group at the end of the previous day, was different in the manner in which objects were counted. One month earlier, during the design of the activities with the teacher (there will be more about co-designing activities in the next chapter), we decided that it would be important to encourage several ways of perceiving the mathematical structure behind the sequence. With this idea in mind, the teacher resorted to a method that was devised by another group of students. The method consisted of conceiving of the terms as being divided into two rows, and counting separately the dark square. As one of the students put it when dealing with Term 6, "We go 6 + 6 equal 12, plus one."

As on the previous day, the teacher illustrated the method through a complex use of gestures, words, and rhythm:

Teacher: (*Pointing to the number of the term*) Term 1; (*Pointing to the bottom line*) one on the bottom; (*Pointing to the top*) one on top; (*Pointing to the dark square*) plus one.

Joined by the students, she counted in the same rhythmic way the other terms up to Term 5:

Teacher: (*Pointing to the number of the term; see Figure 29, Picture 1*) Term 5; (*Pointing to the bottom line; see Picture 2*) one on the bottom; (*Pointing to the

FIGURE 29 The teacher and the students counting rhythmically say (Picture 1) "Term 5," (Picture 2) "5 on the bottom," (Picture 3) "5 on top," (Picture 4) "plus 1"

top; see Picture 3) one on top; (*Pointing to the dark square; see Picture 4*) plus one.

The students were now able to tackle the activity's subsequent questions. Among these questions the students had to find out the number of squares in terms that were not perceptually accessible, such as Term 12 and Term 25. Here is an excerpt from the dialogue of Sandra's group as they discuss without the teacher:

- Sandra: (*Referring to Term 12*) 12 plus 12, plus 1.
- Carla: (*Using a calculator*) 12 plus 12 … plus 1 equal to …
- James: (*Interrupting*) 25. Sandra: Yeah!
- Carla (*Looking at the calculator*) 25!

Then, a few minutes later, dealing with Term 25, Carla quickly says, "25 + 25 + 1 equals 51."

This example shows, as the one discussed in the previous chapter, a semiotic contraction. We see how choices are being made between what counts as *relevant* and *irrelevant*, resulting in a more refined linkage of semiotic resources. It also shows the students coming to inhabit a mathematical concept.

8 Synthesis

In this chapter I presented an overview of embodiment in the theory of objectification and the ensuing concept of sensuous cognition. The perspective outlined here differs from empiricist accounts of embodiment that consider the senses as already given entities and assume that the objects of touch, perception, etc. are given to the body and the senses in their immediacy. These accounts prolong the Western epistemological tradition where the senses and the objects of sensation appear as ahistorical and acultural entities.

Sensuous cognition emphasises the idea that our thinking, feelings, deeds, and in fact all our relations to the world (hearing, perceiving, smelling, sensing, etc.), are *intertwines* of our body and material and ideational culture (de Freitas & Sinclair, 2013, 2014; Malafouris, 2004; Malafouris & Renfrew, 2010). However, it understands the intertwines as *cultural and historical*. It stresses the importance of the cultural and historical mediations in our dealings with the world. As a result, instead of being things independent of our historical and cultural settings, our senses, the objects that our senses feel, and their relationship, change historically and culturally. They change with changes in our material and ideational culture (knowledge, spirituality, etc.). Sensuous cognition posits a dialectical relationship between sensed objects and the sensing subjects, something that is accomplished in practical, material, and sensible activity (Radford, forthcoming, in press).

This concept of embodiment sheds light on learning by allowing us to better specify the nature of the encounter at the basis of the concept of objectification. Objectification, I suggested, is an embodied and material encounter with cultural systems of thinking and doing. The embodied nature of this encounter is not an encounter between a given sentient body and a given object. The sensed object does not offer itself to the sentient body in its immediacy, as something simply there, but as something in becoming, in transformation, from potentiality to something tangible whose tangibility is shaped by the body or the sentient bodies that feel and palpate it through concrete activity. In sensing the object, the object is transformed in object of consciousness and thought, transforming thereby not only consciousness but the sensing body and its senses. I discussed in some detail the case of perception.

In the example discussed in this chapter, the objects of touch, perception, etc., are ideational algebraic objects (e.g., variables, covariations, and formulas). These objects are part of a crafted cultural-historical reality, so that what the students' bodies and senses encounter is not pure stuff, pure concrete or ideal matter, but matter already transformed by human concrete activity. In the encounter with these objects, the students' perception was transformed (gestures too, but I focused here on perception).

The generalisation of numerical-geometric sequences like the one discussed in this chapter required the students to perceive the terms of the sequence in a certain way. To attend something in a certain way, and to "intuit" it in a specific form, to use Husserl's concepts, requires a specific intentional act. The object as it appears in consciousness and the intention to perceive it are coterminous. The problem was then to account for that which would allow the students to move from an initial perception of the given terms of the sequence into a more sophisticated one. We saw that such a transformation was crucial to ensuring the students' increased fluency in a complex form of mathematical reasoning vital to their road towards algebra. This transformation, I suggested, is part of the *domestication of the eye*, or—if one so prefers—in educating the eye; that is to say, in converting it into a cultural-theoretical organ of perception.

Gaining fluency in complex mathematical forms of reasoning does not, of course, consist of a mere transmission of knowledge. A superficial look at the previous passage could lead one to think that the teacher was merely "transmitting" a generalising method. A fine-grained video analysis shows that this is not the case. For instance, the sophisticated form of generalisation reached by the students during the activity required the emergence of certain *sensibilities*, such as perceiving the terms in certain ways that facilitate the calculations required to answer questions about Terms 12, 25, 100, or other terms beyond the perceptual sensorial realm.

The *poetic moment* of disclosure of the general structure behind the sequence discussed in this chapter was the result of a joint student-teacher labour. This moment—the event of the thing in consciousness—was much more than a negotiation of meanings and an exchange of points of view. It was rather a Bakhtinian heteroglossic merging of voices, pointing gestures, perceptions, and perspectives. The poetic moment at the heart of the objectification process was made possible by the teacher's and the students' evolving shifts of attention articulated around a complex configuration of semiotic resources—in particular, gestures, language, and rhythm.

Becoming aware of how such resources can be mobilised can be an important element in our understanding of the phenomena accompanying the learning and teaching of mathematics. This awareness might lead in particular to revisiting the teacher-student relationship formed in the act of knowing (Ligozat & Schubauer-Leoni, 2010; Radford, 2014b). The students' accomplished generalisation may not be as sophisticated as a symbolic generalisation such as $2x + 1$, which could be expected from older students. Yet, although our young students' generalisation is not expressed through alphanumeric signs, it is, I want to argue, algebraic in its own way (Radford, 2011a, 2012a). The "formula" through which such a generalisation is expressed is not made up of letters and other signs (e.g., "+"). It is made up of *actions*. It is better to see it as an *embodied formula* that, instead of being expressed through letters, is expressed through actions unfolding in space and time. The transformation of the students' embodied formula into more sophisticated, purely alphanumeric formulas requires a supplementary refinement of the eye (Radford, 2018c).

Acknowledgement

This is a revised and expanded version of Radford, L. (2010). The eye as a theoretician: Seeing structures in generalizing activities. *For the Learning of Mathematics*, *30*(2), 2–7. Revised with permission of the copyright holder.

Notes

1 We have already met the idea of the transcendental "I"; it was discussed in Section 4.3.1 above. It is the transcendental "I" with which Hegel (1991) deals in Section 47 of his *Encyclopaedia*.

2 In *Empiricism and Subjectivity*, Deleuze (1991) explores the kinds of modern subjectivity that arose out of the empiricist trend of the 17th and 18th centuries, a trend that, with Hume, leads to the problem of the secularisation of knowledge.

3 We can see Marx's distinction in the original sentence: "Der Hauptmangel alles bisherigen Materialismus (den Feuerbachschen mit eingerechnet) ist, daß der *Gegenstand*, die Wirklichkeit, Sinnlichkeit nur unter der Form des *Objekts oder der Anschauung* gefaßt wird" (Marx, 1845; Marxist Internet Archives; emphasis added).

4 These results come from an analysis that we conducted using the voice analysis dedicated software *Praat* v. 5.1.04, developed by P. Boersma and Weenink (see www.prat.org).

CHAPTER 7

Task Design: Or Configuring Teaching-Learning Activities

1 Outline

In this chapter we examine in more detail one of the methodological aspects of the theory of objectification—the component Φ associated with the teaching-learning activity. The component Φ was briefly discussed in Chapter 5. Here, we intend to discuss this component from the perspective of *task design*.[1] Let us recall that, in the terminology introduced before, every teaching-learning activity is associated with a *task*, which is constituted of a series of problems that the students are invited to tackle (see Figure 16, Chapter 5). Two important questions that arise in this context are:

- Q1: What guidelines are to be considered in choosing the *mathematical problems* that the students will be encouraged to solve?
- Q2: In what *order* are the problems to be presented to the students?

However, since we cannot think of the mathematical dimension without thinking of the social dimension as well, we need to ask another question:

- Q3: What are the forms of *human collaboration* that can help the students deal with these problems?

This chapter is organised as follows. In the first part, we discuss these three questions. In the second part, we present an example of task design.

2 General Considerations

The task design that we have carried out for more than twenty years has been characterised by the collaboration of the teachers involved in our research projects (e.g., Radford, 2000, 2018c; Radford & Demers, 2004; Radford, Demers, & Miranda, 2009). The activities have been co-designed by the participants of our projects: undergraduate and graduate students, teachers, and researchers. We have identified three elements to carry out our work: (1) general

 | DOI: 10.1163/9789004459663_008

considerations, (2) considerations concerning the mathematical problems, and (3) considerations about the targeted forms of human collaboration.

1. General considerations include:
   a. taking into account what the students already know; and
   b. involving, as far as possible, the use of (concrete, technological, etc.) artefacts.
2. Considerations concerning the mathematical problems include problems that:
   c. are interesting from the students' point of view;
   d. offer the students opportunities to engage with mathematical knowledge at deep levels of conceptualisation;
   e. are organised according to a *conceptual and contextual unity* (more on this below); and
   f. have an *increasing conceptual complexity* (more on this below).
3. Considerations about the targeted forms of human collaboration include the organisation of the classroom in ways that:
   g. encourage critical reflections; and
   h. propitiate a strong interaction between the students, and between the teacher and the students.

Let us now explain these elements.

First, the purpose of taking into consideration what the students already know is to capitalise on knowledge with which the students are already familiar. By resorting to familiar mathematical knowledge, the students will be, in principle, in the position of "immersing" themselves in the activity. Otherwise, it is very likely that some students may just give up and the mathematical activity of the classroom will not achieve the *goals* and the *object* defined by the teacher's pedagogical project (see Figure 16).

Second, the tasks should, as much as possible, appeal to the students' interests. We try to devise mathematical problems in a playful or narrative context in which individuals' actions motivate mathematical reflection.

Third, in order to promote a progressive encounter with cultural-historical knowledge, we design tasks that include different *levels of conceptualisation*.

– A first level is associated with a concrete sensual experience; that is, with an experimentation and reflection through the use of concrete materials (e.g., the bingo chips mentioned in previous chapters, where the bingo chips were used to model and solve the problem of money saving in the Piggy Bank problem).
– A second level of conceptualisation involves a theoretical reflection based on the use of concrete objects that could open up possible emerging links

to the meaning of mathematical objects. For example, in the Piggy Bank problem, this level is shown in the way Albert relates the expression "twice" to the two rows of red bingo chips arranged in front of each of the glasses.
- A third level of conceptualisation appears with the manipulation of mathematical symbols with which the students raise the previous experience (e.g., concrete, sensual experience) to another level of consciousness.

It should be noted that these *levels of conceptualisation* do not imply a strict order. On the contrary, they can overlap each other or even appear simultaneously during the students' and teacher's teaching-learning activity.

Fourth, the problems and questions are formulated so that they are *interrelated*. We suggest choosing what we term a *conceptual and contextual unity*, like in the Piggy Bank problem. This can be done through a narrative story (e.g., "For his birthday, Marc receives a piggy bank with one dollar. He saves two dollars each week. At the end of the first week he has three dollars, at the end of the second week he has five dollars, and so on").[2] It is against this background that the problems of the task are formulated. This *conceptual and contextual unity* is "contextual" in that it offers a *context* to endow the problems with meaning. It is "conceptual" in that it allows for an interrelation of the conceptual dimension of the problems. One of its advantages is to bring into play, in the same context, various mathematical objects in order for the students to establish a clear relationship between these objects and their interconnections (see Chapter 3 for the concept of object), thereby avoiding a fragmented image of mathematics.

Fifth, problems are organised according to an *increasing conceptual complexity*: more complex conceptual problems appear later than the less complex ones. The idea is that by resorting to a *conceptual and contextual unity* and an organisation of the problems according to an *increasing conceptual complexity*, the students can navigate in a meaningful manner through increasingly sophisticated levels of conceptualisation. The expected result is that these two elements taken together can facilitate a significant and coherent student encounter with culturally and historically constituted forms of thinking and action. Figure 30 provides a visual picture of task design from the viewpoint of the theory of objectification. The line connecting the problems connects them in terms of a unifying context (the *conceptual and contextual unity*) and their *increasing conceptual complexity*.

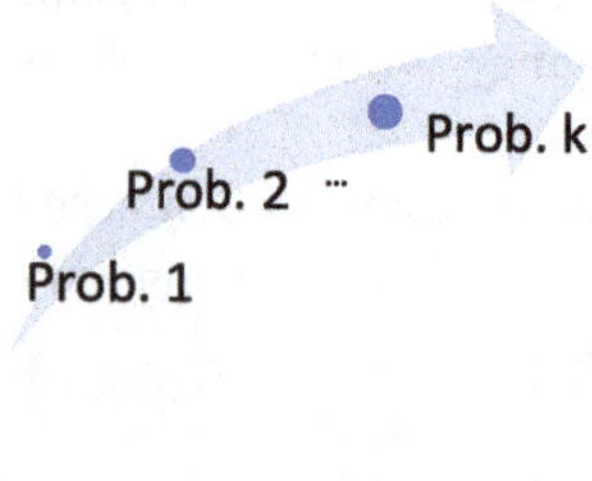

FIGURE 30
The design of the problems based on a conceptual and contextual unity and an increasing conceptual complexity

Sixth, with the task and the social organisation that the teacher proposes, we seek to provoke the encounter with the *Other*; that is, with other voices and other consciousness. In particular, we foster collective forms of knowledge production and modes of human collaboration of a non-alienating nature. The social organisation of the task is about fostering ways and means of interaction that promote critical positions, solidarity, responsibility, and care of others. From the perspective of the theory of objectification, the task, through its epistemological density (shown in the curve of increasing conceptual complexity), should contribute to the creation of a space for reflection, exchange, and debate in which all the people involved, students and teachers, work on what we called in Chapter 2 the production of a *common work*, a work in which the teacher and students unify their efforts and skills and capitalise on their multiple differences and perspectives.

In the next section we discuss an example of task design.

3 The Motion of Tina, John, and the Dog

The following presents an example of task design associated with the production, use, and interpretation of graphs for a Grade 8 class. The example is part of a series of lessons for Grade 7 to Grade 12 students developed in a research project funded by the Ontario Ministry of Education (Radford, Demers, & Miranda, 2009). One of the aims of this project was to provide guidelines for teachers on how to design tasks that require deep levels of mathematical conceptualisation.

The Ontario Grade 8 curriculum outlines the skills that students must develop by the end of the school year. One of these skills included in this official document states that

> By the end of the year the students should use a table of values, a graphical representation, and an algebraic equation to solve problems about relationships [between variables]. (Ministère de l'éducation de l'Ontario, 2005, p. 90)

3.1 *The Object of the Activity*

Taking into account this statement, we proceeded first to define the *object* of the activity. In the terminology that we used in Chapter 5, the *object* of the activity was defined as follows: To create the didactical conditions allowing the students to encounter cultural-historical ways of thinking mathematically about the *motion of objects*. The motion of objects presents indeed an interesting context for graphical and table representation.[3]

3.2 *The Goal of the Activity*

As mentioned previously, in order for the activity to move in the direction of its object, a *goal* should be established. In our example, the *goal* was to get students solving problems requiring the interpretation, production, and use of graphs and tables.

3.3 *The Task of the Activity*

The task should correspond to the goal and the object of the activity. To a large extent, the methodological problem is precisely this: *how to configure the task.*

To design the task, we took into consideration that the contemporary mathematical study of motion involves a very particular way of thinking that mobilises a series of mathematical objects such as space, distance, time, speed, acceleration, Cartesian graphics, equations, and so on. In this order of ideas, the task should allow the student to encounter such a contemporary mathematical way of thinking about motion, which is very different from, for example, Aristotle's way of thinking about motion (see, e.g., Bostock, 1980; Sachs, 2015).

In general, a task consists of a series of problems that, in turn, lead to questions and/or actions. Typical questions start with "What ...?" or "Where ...?" For example: "What is the distance between individual A and individual B when they have walked for 10 seconds? Where is individual A at 10 seconds?"

An *action* to be done is, for example, "Calculate the average speed of A from 0 to 2 seconds" or "Draw a graph that represents the distance between A and B from 0 to 4 seconds."

In the example we want to discuss, our intention was that the problems of the task would lead the students to perform certain key actions in relation to the investigation of motion from a mathematical point of view. These key actions could serve to:

- make connections between the information presented in graphs and the one presented in a table of values,
- compare the information of graphs,
- complete a table of values,
- collect data,
- make sense of graphs through body movements,
- measure time and distance between objects in motion, and
- calculate the value of the (average) speed of an object.[4]

The task that we codesigned with the teachers usually spans several days. In this case it consisted of four problems. Such codesign took into account the forms of human collaboration we want to foster in the activity, which we mentioned before. In this vein, the class was divided into small groups (see Figure 17, Chapter 5). In the course of the activity, the teacher tried to encourage debate and openness to other ideas by means of developing critical positionings, a sense of responsibility and solidarity, and care for others.

The *conceptual and contextual unity* led us to organise the problems in such a way that each problem includes several questions and actions interrelated with other problems. This conceptual and contextual unity corresponds to the motion of three characters: Tina, John, and a dog. John moves away from, or approaches, a fountain at different time intervals. Tina and the dog move away from the same fountain at different points in time.

It is worth noting that not all of the characters appear in the first problem of the task. For example, the dog appears in the second problem. In the second problem, John disappears and reappears in the fourth problem.

The appearance and disappearance of the characters are not naive. Each appearance brings up a series of additional conceptual and procedural mathematical challenges that seek to:

1. reconsider some mathematical objects previously encountered so as to bring them up at deeper levels of conceptualisation than before, and
2. provoke the encounter with new mathematical objects.

We can rewrite Figure 16 as presented in Figure 31.

Figure 31 presents a map of task design. The curved line shown in Figure 30 is what organises the ordered sequence of the four problems on the bottom row of Figure 31.

In the following four subsections, we present each of the four problems of the task.

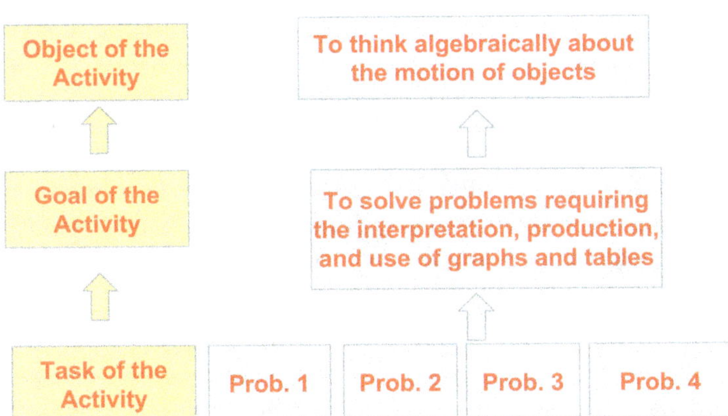

FIGURE 31 The general structure of the component Φ associated with the teaching-learning activity

3.4 *Problem 1: Interpreting and Replicating a Graph*

Problem 1 begins with the verbal description of Tina's and John's positions in relation to a fountain (which is the reference point). Two graphs are shown in the problem, each representing the movement of Tina and John (see Figure 32; a complete description of this problem can be found in the Appendix to this chapter).

1. Tina is 1 metre from a fountain. She starts walking in a straight line. Her motion is represented by the graph on the left.
2. John is 4 metres from a fountain. He starts walking in a straight line. His motion is represented by the graph on the right.

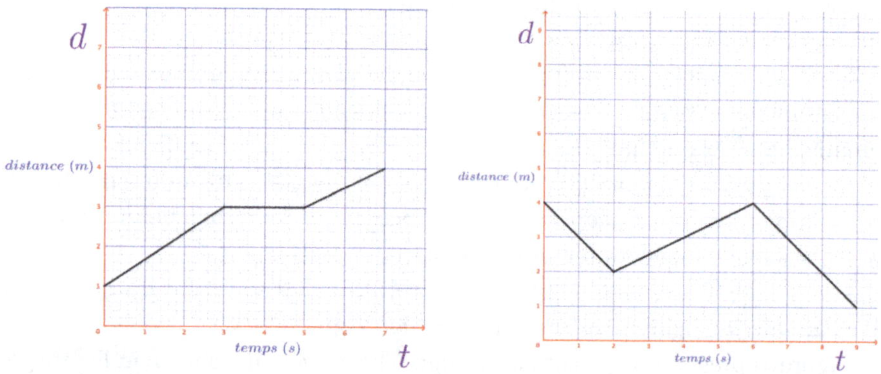

FIGURE 32 Left: The time-distance graph representing Tina's motion. Right: The graph representing John's motion

In this problem, the students are invited to discuss in small groups the interpretation of the graphs and to provide a precise description of it.

First action:

The problem begins with an action (*first action*) to be done by the students. It asks the students to *interpret* the information presented in the graphs. To do so, the students have to *recognise* the different characteristics of the two motions (e.g., moving away from the fountain, moving closer to the fountain, not moving, as well as considerations about the variables time and distance and their relationship).

Second action:

The second action asks the students to provide an accurate mathematical explanation of the graphs. The text in the students' activity sheet for this action is: *Provide a precise explanation (using mathematical terminology) of Tina's walk and John's walk.*

As an example of what can be expected from the students (a response that was foreseen in our *a priori* analysis), a group produced the following text about Tina's motion:

> You must start at 1 metre [from the fountain], take 2 seconds to walk to [the point at] 3 metres [from the fountain], wait for 2 seconds and then walk to [the point at] 4 metres [from the fountain] in 2 seconds. In total, you should have walked for 7 seconds.

As we can see, important points of the walk were indicated according to the distance from the fountain—although, as we expected, the fountain remains implicit. Also, in accordance with previous research (e.g., Radford, Demers, Guzmán, & Cerulli, 2004), the students do not count the elapsed time from the *beginning* of the walk. Instead, they mention the elapsed time for each part of the walk. As a result, although correct from a pragmatic viewpoint, the variable time is not yet conceptualised in Cartesian terms and, consequently, the interpretation is not a Cartesian interpretation of the motion.

Third action:
These two actions are followed by a third action. It consists of asking each small group to replicate Tina's and John's motions. In each small group, the students are equipped with a chronometer, masking tape, and a measuring tape, and are asked to identify a student to replicate Tina's walk and a student to be in charge of replicating John's walk. In each small group, the students are invited to work together.

In general, to replicate the walks, we expect the students to organise the space where the walks are going to occur: they need, for instance, to mark or indicate important points in space. Figure 33 shows one of the small groups taking measurements and reproducing Tina's walk. The three students (Richard, Kent, and Josée; from left to right, left photo) placed labels that indicate key points in Tina's walk (1 metre, 3 metres and 4 metres). The student who was chosen to walk had to reach these key points at specific values of time (0 seconds, 3 seconds, 7 seconds). In the right-hand picture of Figure 33, we see Josée walking, while Richard is measuring time, and Kent, placed at 3 metres from Josée's starting point, is observing her motion.

Once the students are ready, they go into the hallway to join the teacher and try to reproduce Tina's and John's Cartesian graphs.

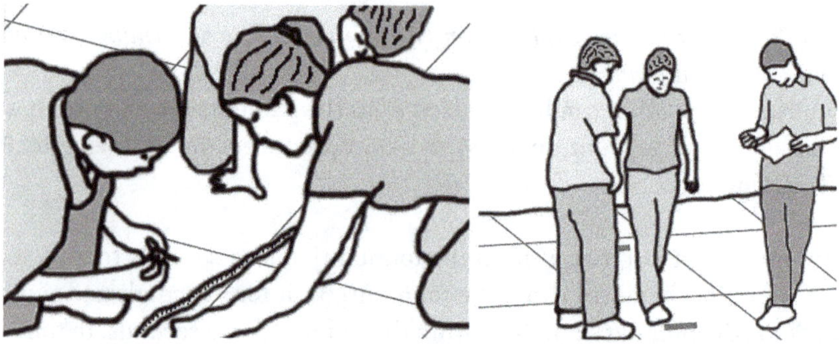

FIGURE 33 Left: the students measuring distances on the floor. Right: the students practising Tina's walk

To reproduce the graphs, the teacher has a motion sensor (called a Calculator Based Ranger or CBR) plugged directly into a graphing calculator (see Figure 34).

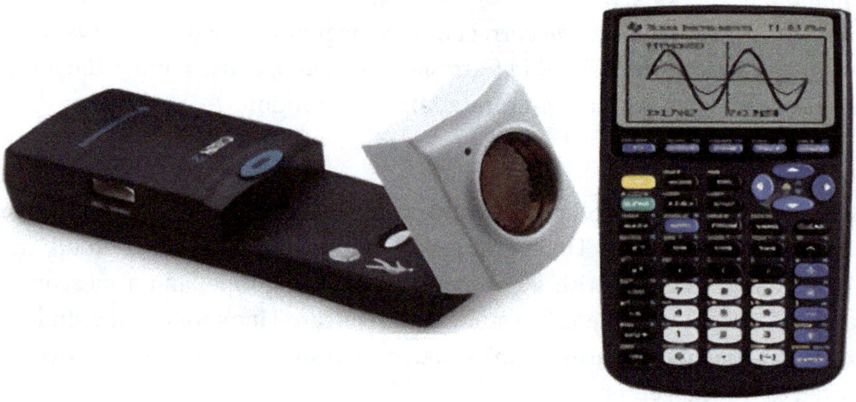

FIGURE 34 The Calculator Based Ranger® (left) and the graphing calculator (right) used by the teacher

The motion uses ultrasound to measure the position of people or objects at short intervals of time (t_1, t_2, t_3, etc.) and records the distance (d_1, d_2, d_3, etc.) of the moving people or objects. Thus, the motion sensor records a sequence of pairs (t_i, d_i) that it sends to the graphing calculator. With the data, the graphing calculator produces Cartesian graphs such as time-distance, time-speed, and time-acceleration.

In the hallway, before the arrival of the students, the teacher chooses a place as the position of the fountain and, on a line in front of the fountain position, puts labels that indicate distances to the fountain (1m, 2m, 3m, and 4m). Then, the teacher gives the starting signal. As the teacher activates the

CBR-calculator system, the student chosen by the teammates to be Tina starts walking (the same is repeated for John's walk). To prevent the students from resorting to trial-and-error procedures, the teacher mentions a condition to be respected at the beginning of the lesson: in the hallway, each group will have only two attempts to replicate Tina's walk (same for John's walk).

After reproducing the walk, the calculator displays the graph. Then, for their records, the students make an approximate copy of this graph into the corresponding Cartesian graph of the problem on their worksheet (see the worksheet in the Appendix of this chapter). Depending on the similarity between the graph produced by the calculator and the given graph in the problem, the teacher *and* the students grade the team's effort (details of this scale are shown in Appendix A of this chapter).

If the graphs displayed by the CBR do not match the ones shown in the problem (see Figure 32), the teacher invites the students to continue thinking and discussing how to reproduce the graphs.

So far, we have described Problem 1 of the task. Let us recall that this problem is framed by a conceptual and contextual unity and is constituted by three specific actions.

The first action asks the students to *interpret* the Cartesian graphs.

The second action asks the students to provide a *mathematical explanation* of the movements of Tina and John.

The third and final action asks students to *reproduce* the movements in front of the teacher. Figure 35 summarises the elements that constitute Problem 1.

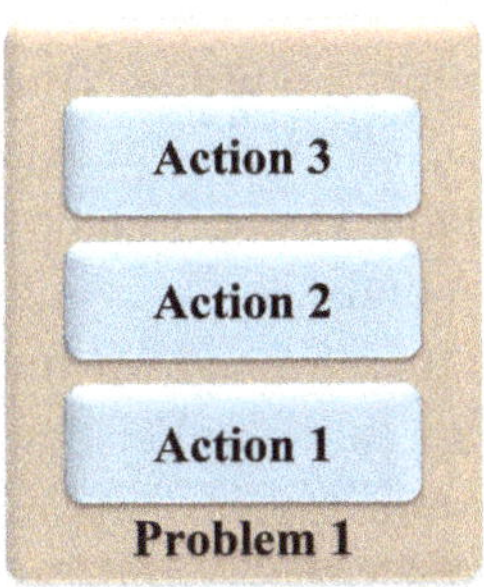

FIGURE 35
Problem 1 and its three actions

Once all the small groups have had an opportunity to reproduce the walks, the teacher may conduct a general discussion and compare interpretations, explanations, and results.

3.5 *Problem 2: Tina and the Dog*

Like all the problems of the task, Problem 2 is framed by the same conceptual and contextual unity. A new character—a dog—is introduced as part of this unity.

First action:
Based on the information of Tina's Cartesian graph, the *first action* of this problem asks the students to calculate Tina's (average) speed during the first three seconds. The students should take into account the concept of speed and establish quantitative relations between distance and time.

First question:
The first action is followed by the *first question*: When does Tina walk the fastest? With this question the students are required to make comparisons between the speeds associated with each of the three segments shown in Tina's Cartesian graph (see Figure 32).

It is from this question that Problem 2 calls for the use of other mathematical objects that are closely related to motion, such as speed, and allows the students to become aware of other attributes that characterise it, thus leading the students to think of motion through their graphical representation at higher levels of complexity than those required by problem number one.

Second question:
Once the first action is done and the first question is answered, the students tackle the *second question*. The question is: "When is Tina at 3 metres from the fountain?" The students have to use mathematical arguments to answer this question. The rationale behind this question is to try to make the students aware, at a deeper level of consciousness, of the relationship between the basic mathematical variables involved: distance and time. In specific terms, the intention is that the students recognise that, between t=3 and t=5, Tina will always be at 3 metres from the fountain (see Figure 32, left). Here, there is an infinite number of solutions. Furthermore, they are named in accordance with the *origin* of the motion in question (e.g., at t=3 seconds, 3.01 seconds, 3.9 seconds, etc. as opposed to time thought of as *elapsed* time from an *ad hoc* point, like in the example cited above where time appears as a number that expresses the time spent in a part of the walk). The way time appears here in the answer to the question allows the teacher to discuss time with the students as *a Cartesian variable*. As we can see, the task intends to provide the students with an opportunity to *move forward* on *the curve of increasing conceptual complexity*.

It is at this point in the development of the story or narrative of the problem that the dog appears. The inclusion of the dog allows us to ask even more interesting and more complex questions. It appears in a brief text that narrates the characteristics of its motion. The text is as follows:

A dog leaves the fountain in Tina's direction at the same time as Tina starts walking. The dog runs at a speed of 0.75 m/s.

Second action:
At this point, the students are invited to tackle the *second action* of problem 2. This action asks the students to fill in a table of values (see Figure 36).

Time *t* (s)	Distance (m)
0	
1	
2	
3	
3.5	

FIGURE 36
Table of values that relates the distance travelled by the dog at certain values of the variable time

The purpose of this action is for the students to calculate the distance between the dog and the fountain by taking into account the dog's speed. It can be seen that, whereas in the previous questions the students have to calculate speeds from the information of the Cartesian graphs, in this question the students have to do the opposite action: to find the distance with the help of the speed.

It is worth mentioning that, as shown in the table, the majority of the numbers of the time variable are integers. The reason we chose one decimal number in the values of the time column was to encourage students to: (1) realise that graphs such as those dealt with here may also include decimal numbers, (2) reflect mathematically at a higher level of increasing conceptual complexity, and (3) do more complex actions than the ones done in the previous questions.

The *a priori* analysis suggested that students may initially choose iterative sums to determine the values of the distance column. Such strategy must be refined to calculate the value of the distance when the variable time includes decimal values. To do so, arithmetic operations such as multiplication or division are required. In brief, calculations with 3.5 seconds move the students forward on the curve of increasing conceptual complexity, for the forms of reflection, action, and expression used by students to carry out this action are more sophisticated and profound than those used in iterated sums. Once the value of the distance corresponding to the value of 3.5 seconds is found, the forms of reflection, action, and expression can be extended to calculate distances associated with other values of time (e.g., 4.5, 5.5, 6.5, 7.5, etc.).

To sum up, Problem 2 includes a new moving character that allowed us to incorporate new actions that involve deeper forms of reflection and actions. Let us recall that this problem started with one action. The *first action* asked the students to look at Tina's Cartesian graph and find out Tina's (average) speed during the first three seconds of her motion. The *first question* asked the students to find out "when" Tina walks the fastest. Here the answer is based on a use of the variables involved and might lead, with the participation of the teacher, to realise that the solution can be expressed as a time interval.

The *second question* is also a "when" question. It asks when is Tina 3 metres from the fountain. Again, the *a priori* analysis suggested that the students would not necessarily consider the variable time as a continuous variable, and that the results could suggest one or three solutions of t; for instance, t=3, 4, 5; or t=3. We observed these answers in our Grade 8 class. An example of a "t=3" answer is the following: "When she [Tina] starts her walk she is at 1 metre from the fountain, then she walks another 3 so she is at 3 metres from the fountain in 3 seconds." The student walks us through Tina's motion, telling us where Tina was at the beginning of her walk. Then the answer follows a kind of imprecise account that culminates with the correct fact that Tina was at 3 metres from the fountain in 3 seconds. But the answer stops there. A more synthetic view of the graph and the variables is perhaps required. We include this answer because we think that answers like this are valuable to take into consideration in order for the teacher to labour together with the students to come up with narratives that would make the students become aware of the infinity of solutions to this question and how they are described mathematically.

Problem 2 ends with a *second action:* to complete a table of values whose information is related to the dog's motion and requires that students take into consideration decimal numbers.

The various elements of Problem 2 (and this is true of the problems that follow) reflect one of the key aspects of the configuration of the task of the teaching-learning activity. They seek to provoke, with the participation of the teacher, the emergence or transition between three important levels of conceptualisation: the concrete sensual experience, the qualitative interpretations, and the quantitative interpretations associated with the use of symbols. Altogether, these levels of conceptualisation seek to enable the students to become aware of, through increasingly deep and sophisticated levels of generality, the mathematical objects involved in the problem.

Figure 37 summarises the elements of Problem 2.

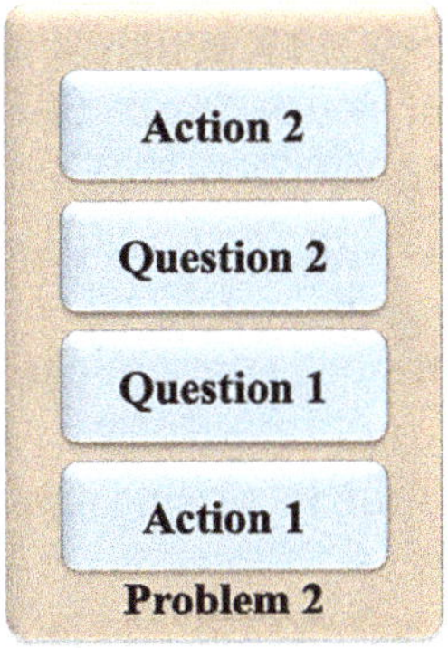

FIGURE 37
Problem 2 consisted of 2 actions and 2 questions

In this problem and the following ones, depending on how the teaching-learning activity unfolds, the teacher decides whether or not to conduct a general discussion about the problem to encourage the students to compare different solutions, interpretations, explanations, and results.

3.6 *Problem 3: The Distance between Tina and the Dog*

With each action and question of this problem we seek to provide occasions for the students to reflect on the distance between Tina and the dog as time goes by. How does the distance between Tina and the dog vary over time?

First action:
The *first action* of Problem 3 required the students to make the Cartesian graph that represents the distance between the dog and the fountain. This action can be done with ease, for the students have previously completed the table of values as seen in Figure 36. Notice that the table of values is an input to do a new action.

Second action:
Once the students make the graph, we ask them to calculate the distance between Tina and the dog at t = 1, t = 2, t = 3, t = 3.5, and t = 4 seconds. This request is the *second action* of Problem 3. It requires the students to observe how this distance decreases. Here the students have to think about generating a time-distance table for Tina or reading the values of the distance from Tina's graph. There is, of course, a conceptual complexity added to this action, for the students have to coordinate different mathematical objects.

Third action:
Then, we ask the students to make the Cartesian graph that represents the distance between Tina and the dog between 0 and 4 seconds. This is the *third action*. We would like the students to visualise how the distance between these two moving characters decreases. The inputs to execute this action have been considered before, for the students already have produced the graph of the dog's motion, they have filled the table of values from Figure 36, and calculated the distance between Tina and the dog. In this way, we encourage the students to use these three inputs in a coordinated way.

It is worth highlighting the way in which the curve of the increasing conceptual complexity of the problems is organised (see Figure 30). Let us note that, in this task, to solve more complex problems, it is necessary to articulate the solutions of previous problems. In this regard, the different solutions of the previous problems are the inputs required to provide the solutions of more complex problems. However, we do not expect the students' recognition of the inputs to emerge immediately; the teacher plays another indispensable role here, since, through the questions he/she asks, he/she will create the conditions for the students to reach deep levels of mathematical consciousness.

On the one hand, through the articulation of the solutions of the problems, we anticipate that the students will see interrelations between the various mathematical objects involved in the elementary mathematical reflection of motion. We expect (and hope!) that the students will come to conceive of mathematical objects as connected between them.

On the other hand, we seek to make the students progressively aware of mathematical objects at increasingly sophisticated and profound levels of generality. This is expressed through the connections that the students can establish between the different components of the mathematical objects involved (e.g., the Cartesian graph, the mathematical origin (0, 0), time, distance, speed, etc.).

First question:
After making the graph that describes the distance between Tina and the dog, we ask the students the *first question*: "What is the distance between Tina and the dog at 1.5 seconds?" The question also asks the students to tell how they found the answer. The idea is that the students use what they have just produced. A graph should serve to do something or enable one to ask questions.

Second question:
The *second question* of Problem 3 was both to determine the time t when the dog reaches Tina, and also to calculate how far away they are from the

fountain. With this question we wanted the students to interpret the point (t, 0) in the graph and to articulate the information presented separately in the graphs concerning the distance between Tina and the fountain and the distance between the dog and the fountain. Without the information from these two graphs, it is not possible to determine the distance between each moving character and the fountain. The reason for this is that the graph representing the distance between Tina and the dog does not involve the fountain explicitly.

Let us summarise what we have said about Problem 3. This problem has three closely related actions. It was configured in such a way that the implementation of some actions of the previous problems served as inputs to solve the two questions of this problem.

The first action of Problem 3 prompts the students to draw the Cartesian graph of the dog's motion.

The second action requires that the students calculate the distance between Tina and the dog at certain values of *t*.

The third and final action asks the students to produce the Cartesian graph that shows the distance between Tina and the dog for t between 0 and 4 seconds.

The first of the two questions of this problem asks to find the distance between Tina and the dog after 1.5 seconds. The second question consists of two parts: When does the dog reach Tina? How far from the fountain are Tina and the dog?

Figure 38 summarises the elements of Problem 3.

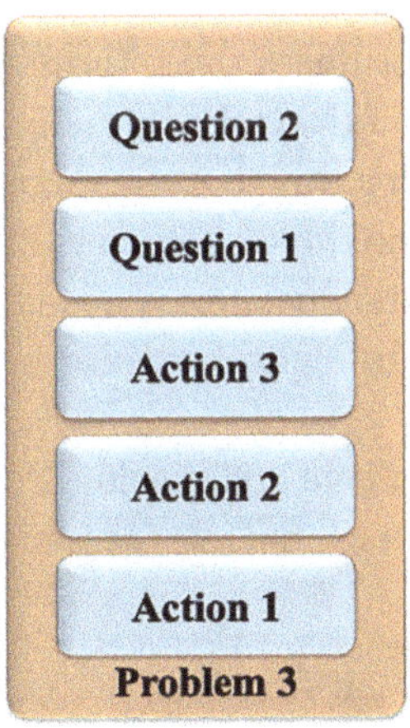

FIGURE 38
Problem 3 consisted of 3 actions and 2 questions

3.7 *Problem 4: A Happy Ending*

The action and questions of Problem 4 involve John, one of the moving characters who appears in Problem 1.

The first action:
The *first action* requires the students to calculate John's speed between 0 and 2 seconds. This action seeks to recapture one of the mathematical objects that had previously appeared in Problem 2. However, unlike Problem 2, in Problem 4, John's distance decreases between 0 and 2 seconds. With this modification, the students are able to visualise similarities and differences between the speeds and distances of each of the three moving characters—even if at this point they do not associate a negative value to the speed (i.e., the calculations do not lead yet to the concept of velocity, which would require elements of Grade 9 mathematics in the Ontario curriculum).

The first question:
The *first question* asks to determine when John is at 3 metres from the fountain. There are three answers to this question. As one of the small groups put it, "John is 3 metres from the fountain at 1, 4, and 7 seconds. This is because he [John] moves away and comes back [from the fountain]. By consulting the graph you can see that he has been 3 metres [from the fountain] 3 times."

The second question:
The *second question* is to determine when John is the farthest from the fountain.

The third question:
The *third and last question* is: If John starts walking at the same time as Tina and the dog, is John far from them when the dog reaches Tina?

The purpose of this question is to have students coordinate the information shown in the previous graphs of the task. This part of the task was called "a happy ending" because the task has been built in such a way that the three moving characters meet each other at the same time. The three characters of the story meet, indeed, after walking for 4 seconds. As the question is posed, however, this meeting may not be evident to the students. Figure 39 presents the answer of one small group.

The English translation is: "John is at the same point as Tina and the dog are because Tina and the dog are 3 m [from the fountain] after 4s; and after 4s John is 3 m from the fountain."

Figure 40 sums up the elements of Problem 4.

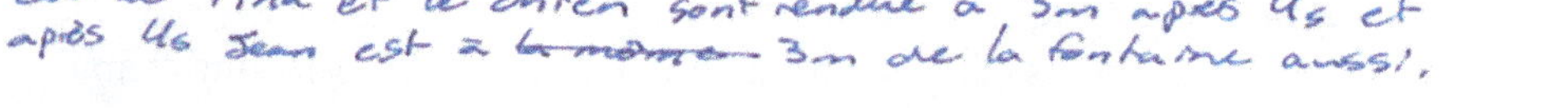

FIGURE 39 A small group's answer to the last question

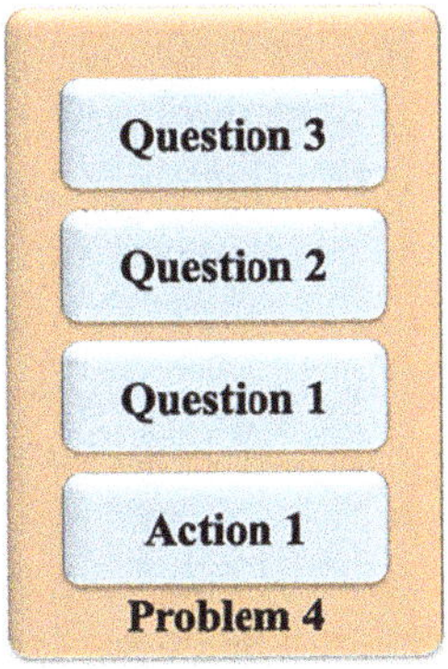

FIGURE 40
Problem 4 consisted of 1 action and 3 questions

Figure 40 summarises the curve of increasing conceptual complexity of the different actions and questions that compose Problems 1, 2, and 4 of the task of the activity (we omit Problem 3 to simplify the curve). The four problems are associated with the production, use, and interpretation of graphs.

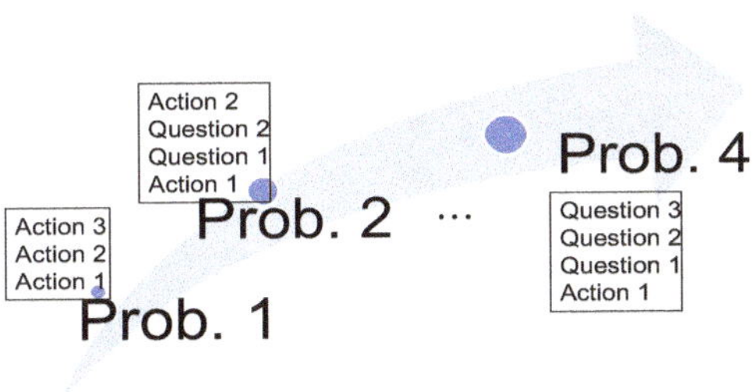

FIGURE 41
The design of the task based on a conceptual and contextual unity and an increasing conceptual complexity

4 Synthesis

In this chapter we have presented one of the methodological aspects of the theory of objectification associated with the task design of the activity that the teacher presents to the students—the component Φ. The main methodological issue revolves around the choice of problems, the sequence of their presentation, and the social organisation of the classroom. The organisation of the task is driven by a *conceptual and contextual unity* that seeks to create deep levels of conceptualisation by means of a curve of *increasing conceptual complexity*.

From the example previously described and its synthesis in Figure 41, we tried to highlight the following aspects.

First, the conceptual and contextual unity is associated with a story that narrates the movement of three moving characters: Tina, John, and a dog. The actions, questions, and their different interrelationships bring together a series of mathematical objects whose particular connections are expected to be encountered by the students through their joint labour with the teacher. The design of the task is expected to help the students to become aware of mathematics as a "coherent whole" composed of individual components and their connections.

Second, the organisation of the problems follows a curve of increasing conceptual complexity; in particular, some of the first problems are used as inputs to solve problems of greater conceptual complexity later on.

The conceptual and contextual unity, as well as the increasing conceptual complexity curve, shapes a teaching-learning activity that should facilitate the students' encounter with cultural-historical knowledge.

Acknowledgement

This chapter was written in collaboration with Oscar Leonardo Pantano Mogollón and Isaias Miranda Viramontes.

Notes

1 In previous works we have used the term "configuration of the activity" because the term "task design" sounds as if the problem at hand is of a very technical nature. However, to conform to a terminology that has become largely used in mathematics education (see, e.g., Watson & Ohtani, 2015), we use the term "task design" throughout this chapter.

2 Jerome Bruner has been a champion in stressing the importance of narratives in human cognition (see, e.g., his book, *Acts of Meaning*, 1990) where he makes a crucial point about the ground narratives offer to organise our experience of the world.

3 Interestingly enough, the *object* of the teaching-learning activity is not part of the mathematics curriculum. Curricula around the world have unfortunately become so technical and precarious in the manner in which mathematics is presented that it is difficult for teachers and mathematics educators to come up with meaningful perspectives to teaching and learning. We have to reinvent contexts that may make meaningful the experience of learning mathematics.

4 The teaching-learning activity discussed here served as the basis for another activity that focused on the production and investigation of equations (see Radford, Demers, & Miranda, 2009, pp. 146–151).

Appendix A: The Walking Competition

Problem 1

1. Tina is 1 metre from a fountain. She starts walking in a straight line. The first graphic below represents her walk.
2. John is 4 metres from the fountain. He starts walking in a straight line. The second graphic below represents his walk.

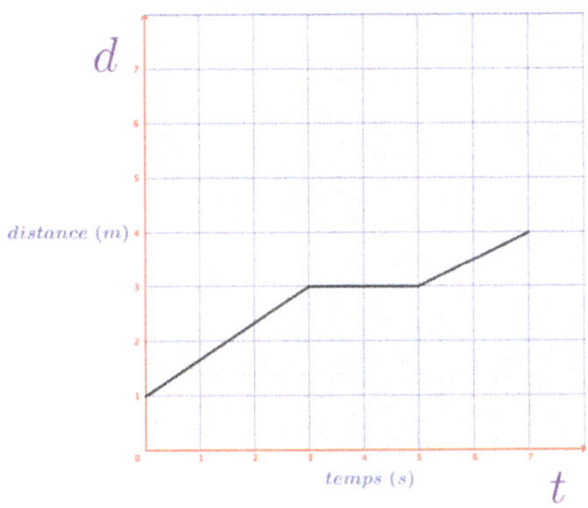

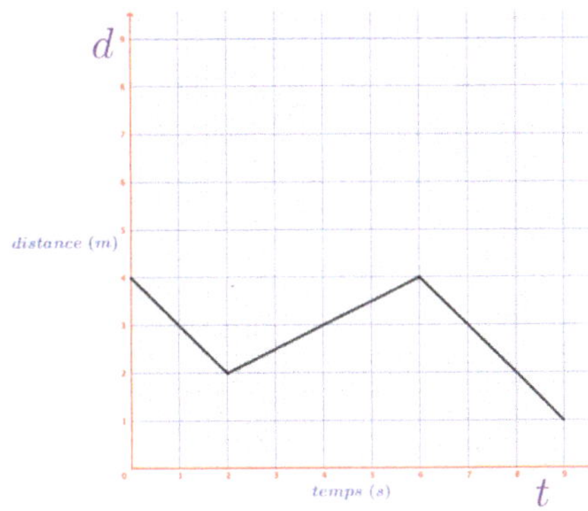

Your group will replicate Tina's and John's walks in the hallway in front of the teacher. Your group can make only two attempts per walk. The points to be won are explained below.

> Points to be won:
> First attempt: A maximum of 12 points (4 points for an exact replication of the first segment + 4 points for an exact replication of the second segment + 4 points for an exact replication of the third segment).
> Second attempt: A maximum of 9 points (3 + 3 + 3).

Before replicating Tina's and John's walks in the hallway, please:

1. Interpret the graphs of Tina and John.
2. Provide an accurate explanation of the walk.

Accurate explanation (using mathematical terminology) of Tina's walk:

____

____

____

Accurate explanation (using mathematical terminology) of John's walk:

____

____

____

3. Practise Tina's and John's walks in the class with your group.

Go with the teacher into the hallway to replicate Tina's and John's walks.

Replicate Tina's walk. After each attempt look at the graph produced by the calculator and make a sketch of your group's replication of Tina's walk:

First attempt

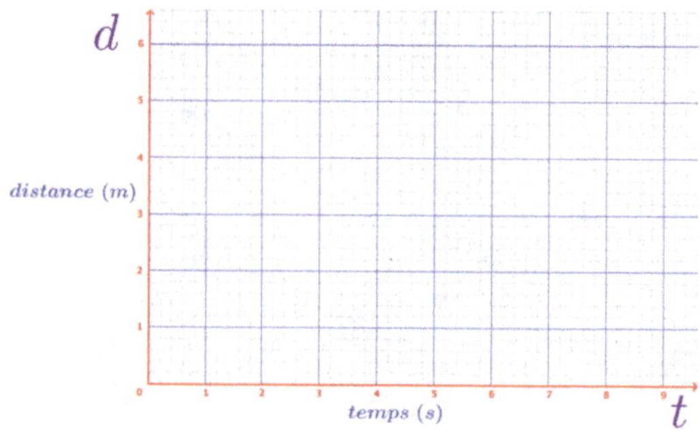

Second attempt

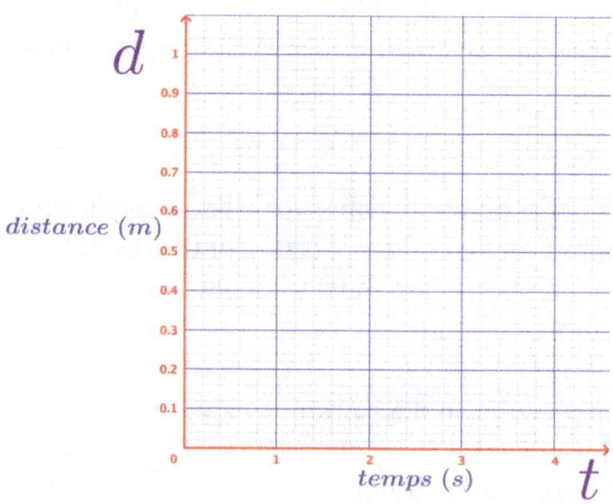

Points obtained by your group:

(To be filled out by your group after discussing/agreeing with the teacher)

Tina's walk: _________

John's walk: _________

Total: ______________

Problem 2: Tina and the Dog

In Problem 1, Tina left at 1 metre from the fountain and walked in a straight line. Here is the graph of Tina's walk.

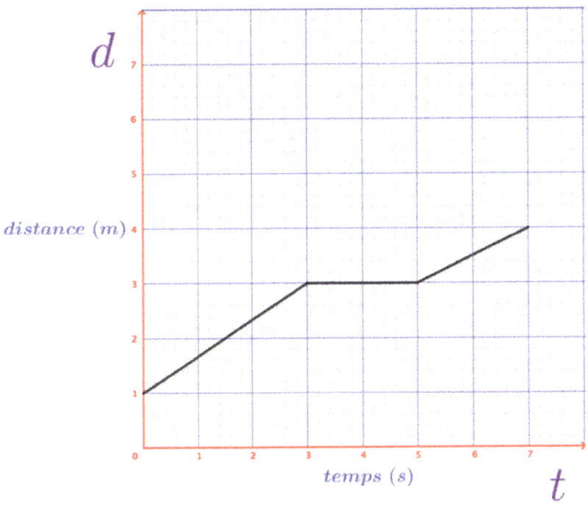

1. Calculate Tina's average speed during the first 3 seconds of her walk. Show your calculations.

2. When does Tina walk the fastest? Explain.

3. When is Tina at 3 metres from the fountain? Justify your answer.

The Dog

4. A dog leaves the fountain in Tina's direction at the same time as Tina starts walking. The dog runs at a speed of 0.75 m/s.

Fill out the table below, which indicates the distance separating the dog from the fountain.

Time *t* (s)	Distance (m)
0	
1	
2	
3	
3.5	

Problem 3

1. Make a graph representing the distance between the dog and the fountain in relation to time. Give a proper title to your graph.

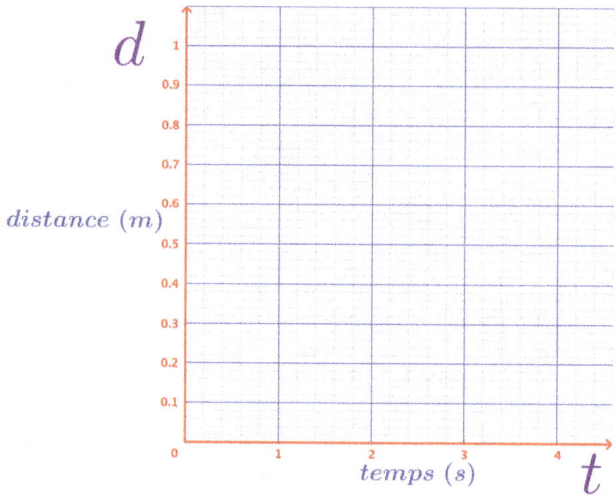

2. Calculate the distance between Tina and the dog when $t = 1$, when $t = 2$, when $t = 3$, when $t = 3.5$ and when $t = 4$ s. Explain your calculations.

3. Make a graph that shows the distance separating Tina and the dog between 0 and 4 seconds, and give the graph a proper title.

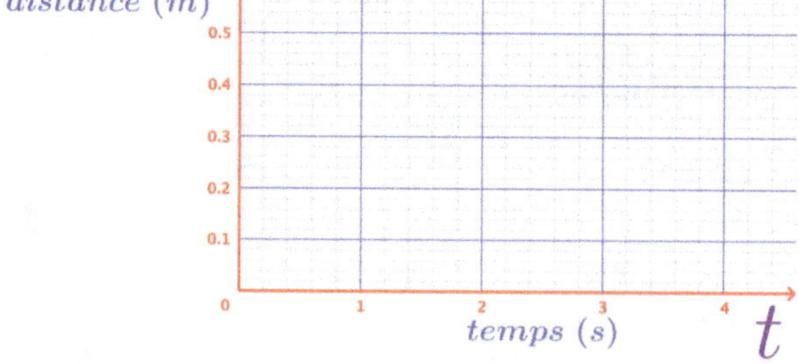

4. What is the distance between Tina and the dog at 1.5 seconds? How did you find the answer?

5. When does the dog reach Tina? How far from the fountain are Tina and the dog? Justify your answer.

Problem 4: A few questions about John and a happy ending

In the previous activity, John was 4 metres from the fountain and walked in a straight line. Here is the graph of John's walk.

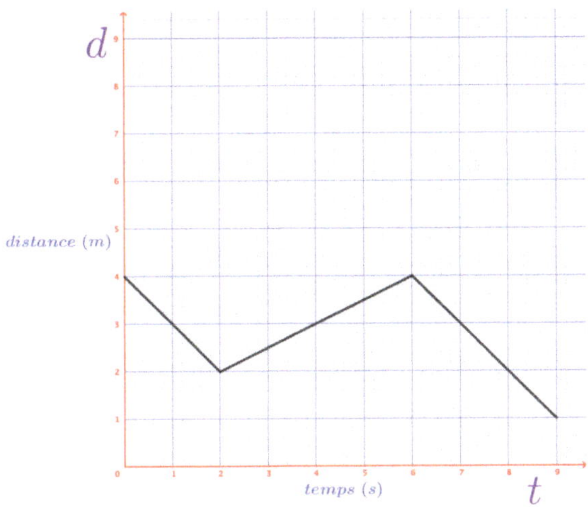

1. Calculate John's average speed between 0 and 2 seconds.

2. When is John 3 metres from the fountain? Explain.

3. When is John the farthest from the fountain?

4. If John starts walking at the same time as Tina and the dog, is John far from them when the dog reaches Tina? Explain your answer.

CHAPTER 8

The Cultural Nature of Mathematical Thinking

> Every culture has a theory that certain "things" actually happen, are "really true," that "have been" or "are"—though, of course, the frames within which these assertions are made may vary from culture to culture. One culture's truth may be another culture's fantasy.
>
> TURNER (1988, p. 41)

1 Outline

This chapter deals with the question of the cultural nature of mathematical thinking. The question is not easy to tackle, as it depends on what we understand by culture. Often, culture is taken as a kind of umbrella that encompasses all kind of processes, thinking included. Here is an example. A few years ago, I was participating in the annual conference of the International Group for the Psychology of Mathematics Education (PME). In one of the research reports, a presenter alluded to the cultural nature of mathematical objects. During the question/answer period, one member of the audience argued that it was not necessary to insist on the question of culture, since obviously everything is cultural. By reducing culture to an abstract overarching umbrella, the problem with this stance is that it becomes impossible to notice the various manners in which cultures affect and shape thinking processes and the ideational objects (e.g., mathematical) that individuals produce. Vygotskian scholar Carl Ratner has shown the limits of this stance that shrinks culture to a mere set of "shared activities, or historically accumulated activities." (Ratner, 2015, p. 57). Such a view "allows virtually anything to qualify as culture, be studied as culture, and used as culture" (p. 57). This abstract view of culture also fails to notice that cultures are heterogeneous sites of conflict and, as such, are bearers of political and economic contradictions of societal processes.

My intention in this chapter is to clarify what is meant by the cultural nature of mathematical thinking in the theory of objectification. In the first part of the chapter, I present an overview of concepts of culture and explanations of cultural differences. As we shall see, cultural differences were first explained

 | DOI: 10.1163/9789004459663_009

through arguments based on an evolutionary view of people and race that used the European as its standard. In the second part of the chapter I present the dialectical materialist concept of culture of the theory of objectification. Through the prism of this concept of culture I look at Greek mathematical thinking to explore the societal contradictions that underpinned it. The chapter ends with an invitation to the readers to turn their eyes to our contemporary forms of mathematical thinking and to reflect on the manner in which these forms of thinking refract and embed the politics and ideologies of our current capitalist forms of production of life.

2 Introduction

In Chapter 2 we noted that sociocultural theories claim that human beings are fundamentally related to the culture to which they belong. To be *fundamentally related* to a culture does not mean to be so in a *contingent* or *accidental* way. The sociocultural claim means that the manners in which individuals think, how they feel and express their feelings, what they do, imagine, hope, or dream are deeply *intertwined* with the culture or cultures in which they live. One consequence is that individuals are profoundly *affected* by their cultures in both what they come to know and the way they come to be.

The nature of this effect of culture on individuals is a contentious matter. It brings to the fore "questions of freedom and determinism, agency and endurance, change and identity, the given and the created" (Eagleton, 2000, p. 8). In individualist approaches, the individual is conceived as a self-made entity. The individual is pictured as if her projects of life, meaning, knowledge, and intentionality come from within herself. The individual is the single author of the life she lives. This is the theoretical conception of the self that emanated from early modernity and received a sharp formulation during the 18th century European Enlightenment movement. In this conception, the social and the cultural are merely by-products of the individuals' actions. Socio-cultural dimensions are mostly seen in contractual terms (i.e., as negotiated obligations between people). While this conception gives the primacy to the individuals, other approaches give the primacy to the cultural; the individual is considered as a product of her culture, shaped either by societal processes of an economic nature or by the symbolic forms of culture. In this chapter, I move away from these two approaches. I explore one in which individuals *create* cultures and cultures *create* individuals. I suggest that the relationship between individuals and culture is *dialectical* in an overdetermined sense to be explained below. In this dialectical relationship, through our engagement in social practices, we affect culture. In turn, culture affects us: it affects both our ways of knowing and thinking as well as our ways of becoming. In this chapter I focus on the epistemological question: the question of culture and our ways of knowing and thinking. In the next chapter I focus on an ontological question: the question of culture and our ways of becoming. Both questions are interrelated. I divide them to simplify the analysis.

As a mathematics educator, my interest does not lie in thinking in general, but in *mathematical* thinking. I am trying to understand what may be called the cultural *nature* of mathematical thinking. A word about my methodology is in order here. We live immersed in cultures—to the extent that much of the cultural dimension in our lives goes unnoticed. It might not be an exaggeration to say that we are a bit in the same situation as a fish that lives in its aquarium. So, methodologically speaking, it might be much better to approach the phenomenon under investigation by turning to cultures in which we are not immersed. There is a phenomenon of *dépaysement* (estrangement, see Guillemette, 2017) that may allow us to better perceive the role of culture in the way we come to think mathematically.

In the first part of this chapter I make a short incursion into anthropology and discuss the concept of culture. In the course of this incursion we shall see that when anthropologists were confronted with the problem of cultural differences in general and differences in thinking in particular, and once they moved beyond racist explanations, they ended up confronted by having to choose between two main options: one that considers the origin of mind and the structure of society in the fulfilment of economic needs, the other that posits rather a symbolic matrix that orders mind and action. I take a different route. Drawing on the work of Eagleton (2000, 2016), Fay (1996), Godelier (2010), R. Williams (2005), and Resnick and Wolff (1982, 2011), in the second part of the chapter, I sketch a dialectical materialist conception of culture. In this perspective cultures are understood as polyglot sites of conflict, resistance, and meaning making embedded and embedding societal processes driven by ideological, economic, and political concerns. In the last part of the chapter I resort to this concept of culture to investigate the *nature* of mathematical thinking in ancient Greece. Conceiving of thinking as the ideational counterpart of social practices, my intention is to offer an explanation about the relationship between the various forms of Greek mathematical thinking in the classical period and the *specificities* of the culture under consideration. In the last part of the chapter I invite the readers to turn their eyes to our contemporary forms of mathematical thinking and to reflect on the manner in which these forms of thinking refract and embed the politics and ideologies of the current capitalist forms of production of life.[1]

3 Boas's Relativist Conception of Culture

People have always recognised that persons from other settings think differently about the world. The ancient Greeks, for instance, referred to the non-Greek-speaking people as barbarians (from βάρβαρος, *bárbaros*), acknowledging that they have other languages, customs, and mores. To the Greek eyes, the Persians, Egyptians, and Phoenicians were barbarians. When, at the end of the Middle Ages, European conquerors, missionaries, diplomats, merchants, and others embarked on their colonising undertakings, they realised that the people they met and subjugated were not thinking in the same way as they did. How were cultural differences understood? For the Greeks it was a question of civilisation versus savagery (hence the pejorative connotation of βάρβαρος). In modern Europe, and until the 19th century, cultural differences were understood in accordance with something that was out of the Greek paradigm of thought: a humanist-evolutionary view of people and race. It was Franz Boas's work at the end of the 19th century that opened new, non-racist avenues to the study of cultural differences and that helped move the discussion to new grounds.

Franz Boas was trained in physics, mathematics, and geography. He left his native Germany in 1883 and embarked on an expedition to Baffin Island in Canada where he carried out research among Indigenous communities to investigate their social, economic, and spatial organisation. In his field studies on perception and language, Boas came to realise that colours, for example, were not invariantly perceived by different cultural groups. With amazement he noted that the Inuit had a rich range of concepts that allowed them to distinguish hues of white. "Here," he wrote, "we find one word expressing 'snow on the ground'; another one, 'falling snow'; a third one, 'drifting snow'; a fourth one, 'a snowdrift'" (Boas, 1944, p. 211). He concluded that the seeing eye is a *cultural organ*. Boas would have not been very surprised to learn that ancient Greeks and Babylonians did not perceive and conceptualise shapes in the same way. Indeed, when dealing with shapes, the Euclidean Greeks saw *surfaces* or *regions*, while the Babylonian scribes saw *sides*.[2]

In the course of his research Boas became one of the most forceful opponents of the idea of biological differences between "civilised" and "primitives." As Cuche (2004, p. 19) notes, "All of Boas' work is an attempt to think the difference ... For him, there is no difference of (biological) 'nature' between primitives and civilized, only differences of culture, acquired therefore and not innate." Stocking summarises Boas's argument against race and primitiveness as follows:

> the general effect of Boas' argument was to show that the behavior of all men, regardless of race or cultural stage, was determined by a traditional body of habitual behavior patterns passed on through what we would now call the enculturative process ... Boas had started out to show that the mind of the dark-skinned primitive shared with that of the white-skinned European the powers of abstraction, inhibition, and choice. He ended by arguing that thought, action, and choice, whether primitive or civilized, were largely determined by the particular body of tradition and custom "that has been controlling all our actions since the time of our birth." (Stocking, 1966, pp. 877–878)

On his return from Baffin Island, Boas tried to find a job in the USA, but a job offer did not materialise. He returned to Germany for a few months and then came back to Canada to embark this time on a field trip to the British Columbia coast to study several communities.[3]

His book, *The Mind of Primitive Man,* is a result of what he learned on his field trips. In it, Boas introduced a relativist conception of culture giving a new direction to anthropological studies. Culture, he says,

> may be defined as the totality of the mental and physical reactions and activities that characterize the behavior of the individuals composing a social group collectively and individually in relation to their natural environment, to other groups, to members of the group itself and of each individual to himself. It also includes the products of these activities and their role in the life of the groups. (Boas, 1944, p. 159)

The relativist conception of culture that emerged from Boas's work led to a reformulation of the question about cultural differences and the question of the relationship between culture and mind. It led to understand rationality as locally situated or—to use the anthropological parlance of the time—that thought and behaviour are culturally *determined.*

But what is it that makes the Inuit *see* the world as they do and to see shades of white where we do not? What is it that led the Greeks to see surfaces where the Babylonians saw sides? What is it that in the end led the Greeks to a form of mathematical thinking and the Babylonians to a different one? In short, how can we explain that thought and behaviour are culturally *determined*? The question of "determination" led the anthropologists to what they call the "venerable conflict."

4 The Anthropological Venerable Conflict

The conflict rests on two major alternative explanations of mind and culture: one based on a utilitarian view of human action (the "utilitarian explanation"), and the other based on the premise of a symbolic order ("the cultural account"). The former stresses the individuals' practical needs, while the latter stresses a conventional symbolic order through which individuals of a culture make sense of their world. As the American anthropologist, Marshall Sahlins (1976), explains,

> The alternatives in this venerable conflict between utilitarianism and a cultural account may be broadly phrased as follows: whether the cultural order is to be conceived as the codification of man's [sic] actual purposeful and pragmatic action; or whether, conversely, human action in the world is to be understood as mediated by the cultural design, which gives order at once to practical experience, customary practice, and the relationship between the two. (p. 55)

The venerable anthropological conflict has at its root nothing more and nothing less than the question concerning human nature. Is this nature primarily economic or symbolic? How should we understand our species? As *homo economicus* or *homo symbolicus*? If you think that our basic impulses are driven to fulfil our needs to survive and to produce things to survive, and that social institutions and their meanings derive from those impulses, you would be in agreement with the *homo economicus* thesis. If, by contrast, you think that our drives to survive are satisfied in ways already shaped by a symbolic order, then you would be in agreement with the *homo symbolicus* thesis. Take the example of kinship of the Tallensi in the French Ivory Coast and the Northern Territories of the Gold Coast. Meyer Fortes saw in this African kinship organisation "a system of values, a body of rules of conduct, and a scheme of social relations" organising society as a whole, including its economic sphere (Fortes, 1949, p. 339). For Peter Maurice Worsley, who defended the *homo economicus* thesis (or utilitarian thesis), kinship is a *derivative* symbolic entity "arising from the needs of agriculture, the inheritance of property, etc.," so that "far from being basic, [kinship] is secondary" (1956, p. 62).

In opposition to the pragmatic reasons invoked by utilitarian theories, Sahlins claims that the ultimate reason lying at the heart of society as its organising principle is "the symbolic or meaningful" (Sahlins, 1976, p. viii). Sahlins concedes that the individuals must live in a material world and that they have to conform to the material constraints they find in their environment, but his point is that the individuals do so "according to a definite symbolic scheme" and that it is this scheme that defines utility (Sahlins, 1976, p. viii). For Sahlins, the sin of the utilitarian approach is to reduce culture to an instrumental system. The utilitarian approaches have not "been able to exploit fully the anthropological discovery that the creation of meaning is the distinguishing and constituting quality of men [sic]" (Sahlins, 1976, p. 102).

Thus, to come back to our mathematical example, it would be futile to look for the particularities of ancient Greek mathematical thought in the Greek forms of economic production, their social division of labour, the appropriation and distribution of economic surplus, the political structure of the state, and the forms of social exclusion. All these aspects of ancient Greek society, including the ones pertaining to the realm of ideas (Greek mathematical thought included) are results of the mediation that a *symbolic order* has already operated. This symbolic order includes the pursuit of truth, the theoretical distinction between being and not-being, the principle of the third-excluded, the Greek standard of perfection, the power of *logos*, and the repugnance for manual labour and the concomitant passion for abstraction. Likewise, it would be futile to look for the particularities of Babylonian mathematical thought in the economic, social, and political structures of the city and administrative needs of the Temple and military armies. The particularities of thought, mathematical and other, would rather be explained by culture itself: "not merely mediating the human relation to the world by a social logic of significance, but constituting by [a cultural] scheme the relevant subjective and objective terms of the relationship" (Sahlins, 1976, p. x).

The *homo economicus* thesis seems to present us with a very reductive view of the species. But so does the *homo symbolicus* one. In the symbolic conception of culture, society is seen as a kind of *artistic artefact*. It makes action and one's everyday existence an *aesthetic* event. Eagleton, R. Williams, and others have seen in this concept of culture a reaction to the entrenched utilitarianism that came along with industrialisation at the end of the 19th century and the beginning of the 20th century. Eagleton notes that, at this time,

> An industrial order powered by rivalry and acquisitiveness has sundered the traditional bonds between individuals, locking each of them into his or her own solitary space. The very fabric of society is consequently in peril. Human relationships have become contractual rather than organic. (Eagleton, 2016, p. 114)

Bringing meaning to the fore and putting it at the origin of action appears as a weapon to counter the disappearance of links and relations that were vanishing with the march towards industrialisation and its emphasis on "utility as the source of value" (Williams, 1953, p. 250). Certainly, the semiotic aesthetic concept of culture articulated by symbolic anthropology overcomes the limitations of the utilitarian concept of culture that derives meaning from the economic sphere. It also overcomes the utilitarian conception of the human as a rational agent seeking self-satisfaction. Yet, the concept of culture in symbolic anthropology as elaborated in the work of Sahlins and others, removes from focus questions of politics, power, marginalisation, and the inherent societal contradictions. It pictures instead a peaceful view of society. Culture can now safely be seen as "an historically transmitted pattern of meanings embodied in symbols, a system of inherited conceptions expressed in symbolic forms by means of which men [sic] communicate, perpetuate, and develop their knowledge about and attitudes toward life" (Geertz, 1973, p. 89). In the end, symbolic anthropology presents us with a sanitised concept of society.[4]

In the next section I would like to explore a concept of culture that takes into account both the economic dimension of society and the fundamental role of meaning in human action—one that is neither a mere reflection of economic structures nor the effect of a semiotic scheme. I would like to explore culture as a dynamic entity that is the product of various complex societal processes. In this view culture "incorporates questions of power and the dialectical construction of meaning"; it becomes a historical entity where "multiple and contradictory positions are constituted" (Simon & Dippo, 1986, p. 198).

5 A Dialectical Materialist View of Culture

5.1 *The Problem of Determination*

In anthropology, the problem of determination consists of first considering society as roughly divided into two spheres, an economic and an ideational or conceptual one; and then considering that the former *determines* the latter. The idea goes back to Marx, who introduced it in his *Contribution to the Critique of Political Economy* (Marx, 1904). The economic sphere or base includes property relations, and intellectual and material forms of production, while the ideational one includes legal, artistic, political, scientific, mathematical, and other systems of ideas. Marx says:

> The general conclusion at which I arrived and which, once reached, continued to serve as the leading thread in my studies, may be briefly

> summed up as follows: In the social production which men carry on they enter into definite relations that are indispensable and independent of their will; these relations of production correspond to a definite stage of development of their material powers of production. The sum total of these relations of production constitutes the economic structure of society—the real foundation, on which rise legal and political superstructures and to which correspond definite forms of social consciousness. The mode of production in material life *determines* the general character of the social, political and spiritual processes of life. (Marx, 1904, p. 11, emphasis added)

A Marxist tradition understood the term "determination" in a *causal* positivist sense and led to vulgar materialism. The economic structure (the so-called "infrastructure") was understood as something unitary of which the ideational (the "superstructure") was its mere refraction. The result was a simplistic view of society and a very impoverished idea of culture.

Raymond Williams (1953, 2005), Maurice Godelier (2010), Terry Eagleton (2000, 2016), and others have emphasised the need to rethink the infrastructure and its relationship with the superstructure. To begin with, we need to dissipate the idea of base or infrastructure as something unitary. "I would argue," says Raymond Williams in his famous book, *Culture and Materialism*, "that the base [or infrastructure] is the most important concept to look at if we are to understand the realities of cultural processes" (Williams, 2005, p. 33). To succeed, however, we have to move away from static and unitary conceptions of the base and recover Marx's emphasis on the *activities* involved in the infrastructure and the contradictions to which such activities lead. Unfortunately, "the base has come to be considered virtually as an object, or in less crude cases, it has been considered in essentially uniform and usually static ways" (p. 33).

We would need to reconsider the base as a site of interactive activities and relationships between real, concrete individuals. The base should be thought of as a site of productive and relational activities in which individuals interact with each other; activities that come into conflict with each other and that, in the midst of those conflicts, contradictions appear keeping the infrastructure and superstructure always in motion. We have, then, to bear in mind that, "when we talk of 'the base', we are talking of a process and not a state" (Williams, 2005, p. 34).

How should we then understand the relationship between infrastructure and superstructure? We are faced here with the exact same problem with which Vygotsky, Leont'ev, and other psychologists of the Russian school of Activity

Theory were confronted when, dealing with the nature of the human mind, they tried to explain the relationship between practical and mental activity. Is mental activity a copy of external, practical activity? (For a discussion on this question, see El'konin, 1995). I touched upon this problem in Chapter 4 when I discussed the controversial concept of internalisation. Sticking here to the question of the infrastructure and superstructure of society, I want to suggest that the answer has to be worked out through a *dialectical* conception of society and a concept of "determination" broader than the causal one of vulgar materialism. The answer needs to be based on a *processual* concept of society, an *agentic* concept of the individuals (that Marx tirelessly brings to the fore in, for instance, his *Theses on Feuerbach*), and a vision of culture in which infrastructure and superstructure are not thought of spatially, but in terms of the interactions and mutual effects within an encompassing dialectical whole. Within this conception, "social life in its totality [is] a complex web of relations in which all factors are interconnected" (Dupré, 1983, p. 88). Indeed, a culture is not made up of isolated pieces. It is a moving whole whose components are interlinked by a dynamic set of societal relations acting simultaneously. Their simultaneity does not imply, however, that these relationships always operate in the same way in the processes of material and spiritual production of social and individual life. The relationships create tensions and conflicts that are always contextual, keeping the culture in perpetual motion. This is precisely the dialectical-materialistic conception of culture that does not give primacy to either materiality or the world of ideas, but conceives of them as two sides of the same coin, as parts of the same process. The ideal (Ilyenkov, 1977) or *l'idéel* (Godelier, 2010), that is to say the ideality that thought and action produce, "is not opposed to the material, because to think is to put matter in movement" (Godelier, 2010, p. 199). It is to put the brain, the kinesthetic gesture, perception, language, tactility, and material culture in movement. Hence, the ideal (and the whole superstructure more generally) is much more than something mental: the ideal is a processual reality that is sensible and non-sensible at the same time.

5.2 *Overdetermination*

To overcome the concept of determination in its causal positivist sense, Resnick and Wolff (1982, 2011) draw on the work of Althusser (2005). They suggest the concept of *overdetermination*. The starting point is a dynamic view of society where different processes are occurring (processes that are, e.g., economic, political, artistic, legal, ethical, scientific, or mathematical in nature). While they occur, these processes influence and alter each other. The concept of overdetermination specifies

> that every social process is a unique site constituted by the interaction of all other social processes in a social formation. No process can be reduced to the effects of a single process or partial subset of processes. Each uniquely determined social process participates in the overdetermination of all other social processes. An overdetermined social process is not a transparency from which each of the constituent social processes can be simply identified and "read." (Amariglio, Resnick, & Wolff, 1988, p. 488)

The overdetermination of social processes "rules out any notion that any one social aspect, such as the economic, can be ultimately determinant or determinant in some last instance of other social aspects" (Resnick & Wolff, 1982, p. 39). In this context, social processes are the result of processes of transformation having different analytical references (political, symbolic, economic), yet bringing together these aspects with different nuances and intensities. We end up here with a concept of culture where art, music, literature, and history are the combined result of the society's various processes. This view of society provides room to see cultural processes as dispersed in society and differentiated among social groups.

In this view, far from being monolithic and homogeneous, cultures appear to be

> inherently polyglot, conflictual, changeable, and open. Cultures involve constant processes of reinscription and of transformation in which their diverse and often opposing repertoires are re-affirmed, transmuted, exported, challenged, resisted, and re-defined. This process is inevitable because it is inherent in what it means for active beings to learn and apply cultural meanings, and in the ideational nature of culture itself. (Fay, 1996, p. 61)

This conception of culture is based on a dialectical effect that the various social processes have on each other. The superstructure (or ideational sphere) is not a *mirror* of the infrastructure, as both keep affecting (and, as a result, transforming) each other mutually, and the societal whole to which they belong.[5]

The dialectical materialist view of culture sketched here does not discard the symbolic world; this symbolic world is repositioned. It is considered to be always present in the shape of each social relationship and the implementation of societal processes, however simple they may be. "There can be no distinctively human activity without signs and values" (Eagleton, 2016, p. 11). Every human action contains or is subsumed in a symbolic dimension that transcends it. The same goes for objects. An object (an amulet, e.g., or a luxury car), cannot be reduced to the materiality that composes it. Around the object there is a symbolic halo that gives it a specific cultural meaning. This symbolic dimension covers not only the concrete objects, but also all thinking processes. Through the symbolic dimension, one acts not only on visible nature as one does with tools, but also on an "invisible power" (Godelier, 2010, p. 197) that drives (in a broad and often subtle sense) the reproduction of life and the satisfaction of needs.

And conversely, each social relationship and the implementation of all societal processes (economic, political, artistic, etc.) create and recreate quotidianly, in their overdetermination, the symbolic. We intersect here the path of symbolic anthropology, although the paths go into different vectorial orientations. In this dialectical materialist account of culture, the symbolic is not at the centre of social processes; nor is it what organises them. There is no centre, only movement. Rather than being the centre, the symbolic becomes entangled with the political and the economic, and it is this entanglement that makes culture what it is.

We are now ready to come back to our question of culture and thinking. I will explore it through the lenses of the sketched dialectical materialist conception of culture. For the reasons explained in the Introduction of this chapter, I will focus on ancient Greek mathematical thinking.

6 Greek Mathematical Thinking Revisited

Like all systems of thinking, mathematical thinking originates in the confluence of various societal processes that, in their interaction, produce and modify each other. As a result, thinking in general, and mathematical thinking in particular, embeds and refracts society's various processes and expresses intrinsic societal tensions and contradictions. From this perspective, it should not come as a surprise that, in opposition to standard views, Greek mathematical thinking was not something unitary. In addition to theoretical or theorematic thinking, exemplified by Euclid's *Elements*, there were, indeed, other types of mathematical thinking. They were associated with different social practices and responded in different ways to the various aspects and processes of Greek society. They can only be identified and understood against the background of Greek society in general.

6.1 *The Applied Geometers' Mathematical Thinking*

Asper (2009) has put into evidence a mathematical tradition of geometer practitioners who were busy with problems of field measurement. To understand their activity and how it was carried out, we need to bear in mind that land (either private or communal) was a cornerstone of Greek identity. Ideally, a Greek citizen was supposed to possess land. Knowing how much land to buy or rent (Jameson, 1978) or how much land to give as a guarantee for a loan were part of economic processes whose conduction required specialised knowledge. The geometer practitioners were a response to those needs. These practitioners were also busy with other geometric calculations required in site constructions, such as determining the number of bricks needed for a given wall.

While the Euclidean tradition seems to have emerged at the end of the 5th century BC (Asper, 2009), the tradition of the geometer practitioners seems to go back to a migration movement from the Near East in the 9th century. The migration movement included craftsmen (e.g., goldsmiths, gem cutters, bronze smiths, and ivory carvers) who travelled to Greek cities "and passed on their skills to the Greeks" (Burkert, 1995, p. 21). The geometer practitioners might have then been part of the manual labourers; that is, those "specialized professionals, paid for their services" (Asper, 2009, p. 123). In all likelihood, they did not have the status of citizens, a status that, as Aristotle claimed, should be reserved for "a particular class of men, to which no one who is constantly engaged in commercial or manual labour can belong" (Saunders, as cited in Aristotle, 1992, p. 183).

In Asper's reconstruction, the geometer practitioners' activity was mainly organised not through written texts but rather an oral practice with a kind of guild structure and education of apprenticeship type. Some geometer practitioners were identified as the "*harpēdonaptai* ('rope-stretchers'), obviously surveyors operating with ropes for measuring purposes" (Asper, 2009, p. 113). Traces of this tradition of practitioners can be found in papyri of later date. Asper gives two examples, the first one is the following:

> Concerning stones and things needed to build a house, you will measure the volume according to the rules of the geometer as follows: the stone has 5 feet everywhere. Make 5×5! It is 25. That is the area of the surface. Make this 5 times concerning the height. It is 125. The stone will have so many feet and is called a cube. (Asper, 2009, p. 110)[6]

The geometer practitioners' mathematical thinking was, hence, non-theoretical. As required by their practice, their mathematical thinking was of a practical nature, based on procedures, without requiring proofs or explanations, much like what we find in Babylonian tablets.

6.2 *Pebble Arithmetic Thinking*

The theorematic and the geometer practitioners' thinking were not the only types of Greek mathematical thinking. There were Greek societal processes that required numerical calculations in business and financial matters, giving rise to a mathematical tradition based on the use of pebbles.[7] This Greek tradition of practical number calculations has not attracted much attention.[8] Yet its investigation is very important for grasping Greek mathematical thinking as a dialectical whole. To understand this tradition—i.e., to understand who its practitioners were, and the societal processes embedding the problems they were working on—we need to turn to one of the main axes of Greek society: the slave.

Slaves in Greece were fundamental to shaping:

a. the ideological and
b. the economic structure of Greek society.

(a) On the ideological side, slaves constituted a defining trait of the concept of citizen. Indeed, to a large extent, the citizen is defined in *oppositional* terms to the slave, outlining in a negation of the latter the boundaries of the personality of the former. Thus, in the model of the Athenian citizen, the citizen owns farmland and slaves. The slave does not. The citizen participates in the administrative, religious, political, and military duties of the *polis*. The slave does not. The citizen possesses judgement and reason. The slave (according to the citizen) does not. In fact, the slave appears as something instrumental. Aristotle, conceiving of the slave as a *tool*, writes: "a slave is a sort of living piece of property" (Aristotle's (1984) *Politics*, 1253 b23, pp. 63–64)—a *living* tool.[9] Slavery was a condition for the existence and exercise of the Greek citizen's freedom. As Jameson put it (1978, p. 122), in Greece, there could have not been "true freedom without true slavery."

(b) On the economic side, slaves with no particular skills carried out domestic duties or tended the master's land; others worked in the silver mines or in their owner's enterprises. Skilled slaves, by contrast, worked closely with their masters in workshops or retail shops, or worked for someone else bringing a wage to their master. Skilled slaves could also run an enterprise for their owner (Jameson, 1978, p. 123). Occasionally, the skilled slaves received a salary from their owner and could save money to purchase their freedom (Zelnick-Abramovitz, 2009).

It is worth noting that the slaves' labour was not only a matter of economic revenue. It also provided the master with the required *time* to engage in the social functions of the *polis* (e.g., ritual, political, and military) and in social activities such as banquets and symposia that were part of the Greek core values and of what was meant to live a "good life."

We can now turn to the practitioners of the tradition of practical number calculations. Where are we going to find them? Often, very often, among the skilled slaves. It is in the economic domain where some of these slaves excelled. More precisely, many of them excelled at *banking*.

Undoubtedly, to a large extent, classical Greece was primarily an agricultural society (Jameson, 1997). The basic unit of Athenian society and economy was the *oikos*—the "household," comprising the family, the family's property, the slaves, and the house. Initially the oikos's economic activities were directed to its own consumption. But by the 4th century BC there was an increase in the cash sales of the oikos's surplus in local markets. At the same time, new maritime ventures along with the production of consumer items in commercial workshops led to the spread of minting money and an increase in the circulation of coins coming from different places (Finley, 1973).[10] It is in this context that, "for the first time, money-based markets gave entrepreneurial businesses a central economic position" (Roberts, 2011, p. 51) and that banks came to play an important role. Certainly, the bank's origin was modest. It is found in a new actor: the *money changer*, who conducted business on the *trapeza*. Trapeza means "table" and refers to the table where currency was exchanged. "Since the new price-denominated, profit-oriented market functioned through these coins, monetary exchange through the trapeza itself involved numerous discrete fleeting transactions, themselves further embodiment of the new economic system" (Cohen, 1997, p. 7).

The rapid evolution of this new aspect of Greek economy led to a range of financial activities and its consolidation into what may be more properly termed *banking activities*. They included "lending directly and through intermediation, protecting collateral, accepting deposits, negotiating damages for failure to protect deposits, investing in real estate, making provisions for continuity of banking businesses, executing purchase orders, safeguarding documents, and so on" (Cohen, 1997, p. 21). Among the best-known bankers are Phormiōn and Pasiōn (to whom I will come back later), and Pylades, Cittus, Heracleides, Xouthos, and Dēmomelēs (see Cohen, 1997; Demosthenes, 2004).

Naturally, like in the surveying practices discussed above, the bank activities required a certain knowledge that was more and more specialised. It was a *technē*, a *practical* knowledge. This is why the banker was professionally considered a craftsman, someone practising a craft or trade, like sculptors did.[11]

A stele from the *Inscriptiones Graecae* collection (IG, I^2 324) (Meritt, 1946) involves loans, the time of the loan, and the profit from the interest. An excerpt is shown in Figure 42. The first column indicates the line number, the second column indicates the principal, the third column shows the number of days since the loan was made, and the fourth column shows the corresponding amount that the interest has earned. "T" stands for talent; the string 1095/5 means 1095 drachmas and 5 obols, etc. A talent was equal to 6,000 drachmai; a drachma was equal to 6 obols.[12]

I.G., I^2, 324	*Principal*	*Days*	*Interest on Stone*
line 103	766T 1095/5	1464	37T 2338/2½
line 85	1T 1748	17	4/2½
line 86	521	17	0/1¾
line 87	80	17	0/½
line 88	3418/1	17	1/5½

FIGURE 42 Some lines from the stele identified as IG, I^2 324 (from Lang, 1964, p. 146)

Now, to conduct the banking activities, bankers needed qualified personnel. However, they could not find them among the citizens for the very simple reason that citizens (which were, by definition, *free men*) were not for hire; Greek citizens were not supposed to work for other citizens; they should dedicate themselves to the social and cultural activities of the *polis*.[13] So, bankers turned to the slaves of their own *oikos*. The bank owner taught the slaves assigned to his bank the *technē* required in the banking activities. This was the case of the banker Pasiōn and his non-Greek slave Phormiōn. Demosthenes tells us that Apollodorus, in the middle of a lawsuit, reminds the jury that his father Pasiōn, "who acquired [Phormiōn,] was a banker, and ... taught him [Phormiōn] to read and write and trained him in the business and put him in control of vast sums of money" (Demosthenes, 2011, p. 261).[14]

This is how a number of skilled slaves became banker functionaries and eventually bank owners. The *polis* might have also employed individuals with the suitable financial *technē* to keep track of interests, repayments, and other transactions related to the city administration.[15]

We do not know exactly how the calculations were carried out. The steles and other historical documents just mention the results (as in IG, I^2 324; see Figure 42). What we do know is that, to calculate, stones or pebbles were used:

> Counters—and not written or spoken symbols—were the medium for the manipulation of numbers in the Greek world ... counters were not some *aid* to the manipulation of number, itself understood primarily in

> other terms. They were the medium of numerical manipulation par excellence, in exactly the same way in which, for us, Arabic numerals are the numerical medium *par excellence*. We imagine numbers as an entity seen on the page; the Greeks imagined them as an entity grasped between the thumb and the finger. (Netz, 2002, p. 329)

Lang (1957, 1964) suggests that the calculations were carried out on an abacus and offers an abacus-based reconstruction of some calculations. Historians have suggested that, originally, the abacus (αβαξ) was a flat surface covered with sand in which lines and figures were drawn (Gow, 2010; Netz, 2002).[16]

To sum up the discussions about the Greek practical numeric tradition, historical records show that there was a range of calculators engaged in banking as well as in bureaucratic administrative activities that were required in the economic sphere of the *polis*. These calculators included some bankers' slaves and perhaps also some slaves of the administrative *polis* as well as some free men (e.g., freed slaves and foreigners). Bankers used to choose the most skilled slaves and passed on the knowledge that was required in the various banking activities. Because the practical numeric tradition was carried out with pebbles, no records of its algorithms are found. The necessary *technē* underpinning this practice was not formal. The calculators' mathematical thinking was not theoretical. Like the applied geometers', this form of thinking was not based on definitions and proofs. It was concrete, unfolding in concrete actions, and organised on improvised or refined abaci (in marble and perhaps wood too), relying heavily on perception, the kinesthetic movement of pebbles, and orality.

It is now time to discuss the best-known tradition of Greek mathematics: the theorematic tradition and its peculiar form of mathematical thinking.

6.3 *Theorematic Mathematical Thinking*

One of the reasons this tradition is better known is because it had recourse to written texts. It came down to us through various ancient papyri. Of course, the fact that this tradition relies on the written as opposed to the other Greek mathematical traditions cannot be taken for granted. It is, as we shall see, already a symptom of its ideological, economic, and political place within the *polis*.

The most notable text of the theorematic tradition is Euclid's *Elements*. Euclid's book *Elements* exhibits a detailed organisation and extraordinary rigor of the content. Axioms are clearly presented from the outset and the mathematical objects are meticulously defined. From the axioms, propositions are asserted and proven deductively. Referring to this tradition, Netz notes a curious fact that unavoidably strikes the readers of the *Elements*: the marginalisation of the numerical:

> This [marginalization] is extremely remarkable for a field that, in all other cultures, is mostly organized around numbers ... actual numbers are practically never mentioned in Euclid's *Elements*, a work that *contains three fundamental books on arithmetic*! ... Greek mathematics is centred on geometry and its qualitative features: Greek angles are not measured (inside Greek geometry proper) by degrees, and Greek lines do not have numerically determinate lengths. (Netz, 2002, p. 346, original emphasis)

Numbers for the theorematic tradition are, indeed, not those entities related to the palpable manipulation of pebbles of the Greek calculators. For the theorematic tradition, numbers are abstract things. It would seem that theorematic mathematics and pebble mathematics came from two different planets. But we will see that this is not the case. In actual fact, these traditions were deeply related.

Another fact that strikes the reader of Euclid's *Elements* is that the theorematic text unfolds against the backdrop of a "rhetoric of impersonality" that intends to "convey objectivity" (Asper, 2009, p. 119). There are no personal pronouns, no explicit addressee:

> With the exception of exactly one formula that serves to introduce the repeated claim and marks the beginning of the proof ('I say that ...'), these texts never introduce an authorial voice nor do they ever address the reader. (Asper, 2009, p. 118)

Let us consider an excerpt of Proposition 34, Book I of Euclid's *Elements* and the accompanying diagram (see Figure 43). The text reads as follows: "In parallelogrammic areas the opposite sides and angles are equal to one another, and the diameter bisects the areas" (Heath, 1908, p. 323). Here is the first part of the proof:

> Let ACDB be a parallelogrammic area, and BC its diameter; I say that the opposite sides and angles of the parallelogram ACDB are equal to one another, and the diameter BC bisects it. For, since AB is parallel to CD, and the straight line BC has fallen upon them, the alternate angles ABC, BCD are equal to one another. (Heath, 1908, pp. 323–324)

Euclid goes on to prove that "the triangle ABC is equal to the triangle DCB."

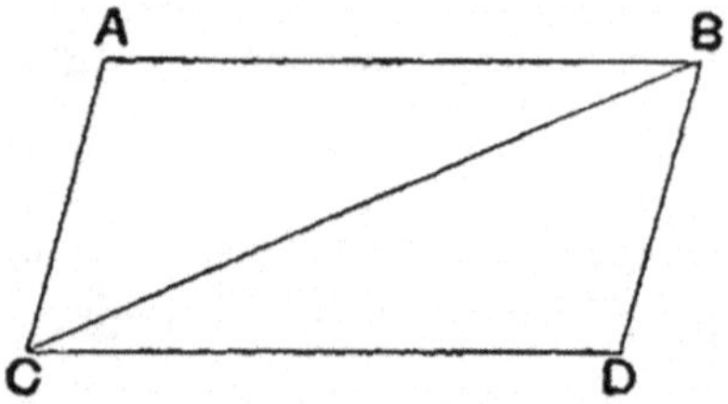

FIGURE 43
The diagram accompanying the proof of Euclid's Book I, Prop. 34 (from Heath, 1908, p. 324)

There is no point in presenting and discussing the whole proof here, as my interest is not in making an exposition of the mathematics involved (for exposition and analysis, see Netz, 1998, 1999). My interest is rather in trying to understand—as I did with the applied geometers' and pebble calculators' traditions—who were its practitioners and the form of thinking of these practitioners in light of the societal processes involved.

The first question to be asked is the following: Who were the practitioners of the theorematic tradition? The answer is not complicated. We do not need to dig deep to find them. It was a social homogenous elite of Greek citizens "free from economic pressures" (Asper, 2009, p. 123)—an elite of those men leading an exemplary "good life," exercising those high faculties praised by Aristotle, Plato, and the aristocratic Athenian society, "notably the [faculty] of reason" (Saunders, as cited in Aristotle, 1992, p. 183). The practitioners of the theorematic tradition were, hence, an elite of an elite: a "social elite who alone wrote and read for pleasure" (Asper, 2009, p. 108).

Theorematic mathematical thinking is the *antithesis* of the mathematical thinking of the practical geometers and calculators. While practical geometers and calculators were measuring fields, calculating areas, volumes, perimeters, and loan interests, the theorematic practitioners were not. While practical geometers and calculators were using ropes, pebbles, and abaci to think mathematically and carry out their calculations, the theorematic practitioners were not. While the practical geometers and calculators were able to go along without proving or defining things, the theorematic practitioners were not. While the mathematical objects of the practical geometers and calculators were concrete, those of the theorematic practitioners were not. While the mathematical thinking of the practical geometers and calculators was orally based, the one of the theorematic tradition was not. One was a literate mathematics, the other not.

In short, when we contrast the two groups, we see that each group produced a mathematics (conceived here as a social practice) and a form of mathematical thinking (conceived here as the ideational counterpart of such a practice). The mathematical way of thinking of one group is the exact negation of the mathematical way of thinking of the other group. The mathematical thinking of the practical geometers and calculators and the mathematical thinking of the theorematic mathematicians refracted and embedded the central contradictions of Greek society, the same contradictions that we found in terms of freedom and slavery: one could not exist without the other. In terms of the dialectical materialist conception of culture previously sketched, those forms of thinking are ideological counterparts of contradictory societal processes overdetermining them.

Now, the nature of their overdetermination makes each form of mathematical thinking something *more* than a simple opposition to the other. To say that they are dialectically (over)produced means that one is the *effect* and the *condition* of existence of the other.

Their overdetermining constitution means that the mathematics of the geometer practitioners and the calculators is *part* of the *texture* of theorematic mathematics. The former is constitutive of the cultural-historical borders of the latter and vice versa. In their refraction of the various Greek societal processes, they are both *embedded* in the essence (to use a Greek word) of the other. This is why it would be a mistake to see the theorematic practitioners' passion and obsession with abstraction as a historical curiosity or the beginning of the unfolding of a Mathematical Spirit (*Mathematischer Geist*) that in its majestic flight seeks perfection through a series of historical periods (Hellenic period, Middle Ages, the Renaissance, the modern period, etc.) spreading its grace and influence more and more to all corners of the earth. There is, indeed, a profound social, political, and economic sense to the shape that Greek theorematic mathematical thinking took—no more and no less than the political and economic sense that underpinned and shaped the pebble and applied geometer mathematical thinking. Thus, the abstraction on which the theorematic tradition is based needs to be understood in its cultural sense. Abstraction (in the aristocratic Aristotelian sense) is a theoretical and ideological device expressing the profound political and economic conflicts of the *polis*. It expresses the fundamental societal oppositions between slave and master, reason and action, possessing or not possessing land, having or not having civic rights. The abstraction of the theorematic tradition is a continuous reminder and proclamation of those divisions. It proclaims through its own exercise the division between the sensual and the ideational, the practical and the rational, and sets the parameters for the division between intellectual and manual labour that still haunts us today.

To better understand the underpinning overdetermination of the societal processes at the basis of Greek mathematical thinking, let me explore further the ideological component that inspires its theorematic mathematicians. I want to focus on the division between the sensual and the ideational. In a passage from Plato's famous dialogue, *Phaedo*, Simmias is asked to determine who, among all sorts of men, is able to attain true knowledge. Is it not he, Socrates asks,

> who pursues the truth by applying his pure and unadulterated thought to the pure and unadulterated object, cutting himself off as much as possible from his eyes and ears and virtually all the rest of his body, as an impediment which by its presence prevents the soul from attaining to truth and clear thinking? (Plato, 1961, 65e–66a, p. 48)

He then continues: "we are in fact convinced that if we are ever to have pure knowledge of anything, we must get rid of the body and contemplate things by themselves with the soul by itself" (Plato, 1961, 66b–67b, p. 49).

With great precision, Plato articulates the epistemic opposition between the sensual and concrete of practical labour, and the disembodied and abstract of intellectual labour. In so doing, Plato and his aristocratic fellows created a segment of Greek culture. Through their practices the pebble and applied geometers created Greek culture too, although perhaps we still need to learn to better appreciate it.

But this is just half of the story. Indeed, we still need to see that what comes out of Plato's mouth is not only Plato's insight; it is not only Plato's incredible and impressive ability to express ideas so neatly. What comes out of Plato's mouth is, overall, the societal tensions of his own city and his own historical period. Plato's argument translates, on the cultural philosophical plane, as Euclid's *Elements* does on the mathematical one, the same societal tensions. In the formidable vocalisation of these tensions, Greek culture creates Plato.

This is why, in the same way as slaves and labourers were admittedly necessary to the life of the city, as Aristotle (1992) tells us in his *Politics* (e.g., p. 180, 1277a5, p. 183, 1277b33), the geometer practitioners and calculators were definitely necessary to the existence of theorematic mathematics. If we remove ropes, pebbles, abaci, etc., if we remove the geometer practitioners and calculators, if we remove the social, political, and economic processes they were part of, we would remove at the same time the *material* and *ideological* support for the theorematic mathematicians to rise and work.

I argued elsewhere (Radford, 2018d) that Plato's ideas about philosophy in general and mathematics in particular arose and evolved during the turmoil of the Peloponnesian War between Athens and Sparta and the post-war oligarchic Athenian regime established by Sparta. The societal processes that I mentioned before—in particular the production of agricultural surplus of the *oikos* (or household), the rise of commerce, and the emergence of new social classes—led, in the aftermath of the Peloponnesian War, to a social restructuring where the old values of the aristocratic elite were shaken. The concept of the "good," related to manliness and good birth, which had been progressively elaborated since Homer's time, was challenged by the new cultural context shaped by the arrival of "[r]ootless foreigners in their origins; skeptical, nominalistic, subjectivistic, and relativistic thinkers ... [who] had no axioms, no epistemic certainties, no fixed axes of value, no ancestral pieties" (Levi, 1974, p. 61). "We can be certain," states Beavers (n.d.) in his biography of Plato, that the Peloponnesian War, "the establishment of a government by Sparta (after the chaos of Athens' final defeat in 404), and the events that followed, dramatically affected the direction of [Plato's] thinking." They affected Plato and the whole aristocratic elite.

Plato grew up in an aristocratic family. His "father's lineage went back to the first kings of Athens" (Levi, 1974, p. 58). Because of his aristocratic ancestry, he was destined to become a member of Athens' ruling class. His path, however, was interrupted by the Peloponnesian War and the subsequent course of events, which led to the decline of the Athenian empire and the Athenian aristocracy's loss of political power. Plato travelled for several years and after his return to Athens in ca. 387 BC, he founded the Academia with the aim "to instruct a new generation to become the legislators and the aristocratic statesmen of a future world" (p. 60). It is against this historical-political backdrop and the aristocratic outlook that opposed epistemological relativism—embodied in particular in the sophists' teaching—and despised social and political change, practical labour, commerce, and all mundane activities, that Plato came to formulate his philosophy of permanent Forms and the ensuing idea of truth as something immutable, perfect, and timeless. In Plato's dialogues, as in Euclid's *Elements*, the objects of true cognition are timeless and immutable objects, which do not come into being and pass away (Burkert, 1985).

Plato's theory of Forms and Euclid's mathematical objects are a nostalgic and reactionary aristocratic move through which truth came to be conceived of as something that was accessible not through practical labour with artefacts but through *logos*, the reasoned discursive activity of cultured citizens whose aim was to rise to higher levels of knowledge and to show to others an example of genuine Athenian "good life."

7 Synthesis

In this chapter I dealt with the question of culture and mind. I focused on the epistemological aspect of this question, more precisely on the *cultural* nature of mathematical thinking. Even if it is recognised that mathematical thinking has not always been the same, that it has changed over time and that it is not necessarily the same from one culture to another, the problem of explaining its cultural nature remains a problem that has not yet been sufficiently studied. In the first part of the chapter, we saw that the problem of cultural differences was first explained through arguments based on an evolutionary view of people and race that used the European as its standard. In his chapter, *The Dark-Skinned Savage: The Image of Primitive Man in Evolutionary Anthropology*, Stocking notes that at the dawn of the 20th century

> Darwinian evolution, evolutionary ethnology, and polygenist race … interacted to support a raciocultural hierarchy in terms of which civilized men, the highest products of social evolution, were large-brained white men, and only large-brained white men, the highest products of organic evolution, were fully civilized. (Stocking, 1982, p. 122)

Boas and others countered this view by emphasising the importance of culture, which he considered as a totality of mental and physical reactions and activities characterising the manner in which people think and act. Culture turned then to be no longer the expression of civilisation (Elias, 2000) nor the result of evolution, but a feature of habits and values nurtured by a community. The question became about that which brings such habits and values into fruition. It was here that we encountered the anthropologists confronted by their "venerable conflict"—the conflict that led them to choose between the utilitarian option that sees our species primarily as *homo economicus* and the symbolic option that sees our species primarily as *homo symbolicus*. Drawing on Eagleton, Godelier, Williams, and Resnick and Wolff, I sketched a concept of culture from a dialectical materialist viewpoint whose first feature is to see society as a dynamic whole where different processes are occurring (processes of, e.g., an economic, political, artistic, legal, ethical, scientific, and mathematical nature) and are affecting and modifying each other. Each societal process is an overdetermination of the totality of societal processes. Culture here appears as a site of confluence, something "inherently polyglot, conflictual, changeable, and open" (Fay, 1996, p. 61). Resorting to this concept of culture, where neither the economic nor the symbolic is taken as the primary ground, I came back to the problem of the cultural nature of mathematical thinking. I focused on ancient Greek mathematical thinking as a case study. To explore its cultural nature, and in accordance with the outlined dialectical materialist concept of culture, the problem was to identify the societal processes at the origin of Greek mathematical thinking. These processes include a focus on land as both an economic and ideologic element of ancient Greek identity, administration needs arising from the organisation of the polis, as well as economic and financial processes of a banking nature entangled with the axiological axis of slave-master and a historical presence of aristocracy challenged at the aftermath of the Peloponnesian War. The overdetermination of these processes resulted in a mutual social awareness of a newly conceived political autonomy of peasants, merchants, commercial bankers, and aristocrats bringing together a new (and always tense) symbolic and legal order "within an emergent sphere that legitimizes difference (and therefore conflict, dialogue, demagogy, reflection, compromise) and thus posits public (third person) justice as constructed rather than given, and necessarily 'compromised' rather than absolute" (Nafissi, 2004, p. 404).

Against the background of these societal processes we were led to distinguish three mathematical thinking forms: first, one related to practices of measuring, and surveying; second, one related to practices of numerical calculations in business and financial matters; and third, one related to a theorematic or theoretical practice. The point, however, was not to make a list of forms of mathematical thinking. The point was to understand their overdetermination. It is in this context that I made an effort to show that one form of thinking cannot be understood without understanding the other forms of thinking, and that such an understanding can only be accomplished by investigating the economic, political, and ideological dimensions of the societal processes that such forms of thinking embed and refract. In this view, one of the distinguishing features of ancient Greek mathematical thinking is that, as a whole, it was produced by social practices that were antagonistic to each other, leading, at the ideational level, to forms of mathematical thinking refracting their antagonism and contradiction. We saw in particular that theorematic mathematical thinking is the exact *negation* of the other Greek concrete forms of thinking. Plato offers us, through his work, an invaluable window to see the dialectical constitution of culture and subject: Plato formulates with incomparable brilliance a view of the *polis*; he brings us to see what a good Athenian is and what to live a good life must be. But at the same time, Plato is an embodiment of the Athenian tensions and contradictions. With all his brilliancy, Plato is a *product* of those tensions. And exactly the same can be said of Euclid.

If we now turn to ourselves and our cultures, what could we say? What are our contemporary forms of mathematical thinking? And how do they refract and embed the involved societal processes of the world we live in? How do our pedagogical practices try to nurture those forms of mathematical thinking?

The increasing multicultural face of schools in the contemporary globalised world cannot remain silent to these questions. In actual fact, there has been an impressive number of research and initiatives that seek to tackle the question of multiculturalism or multilingualism. There has been, in particular, a tendency to become "inclusive" and to appreciate cultural differences. Although commendable in principle, we should however remain vigilant, as behind this noble gesture may rest perhaps the all-encompassing capitalist emphasis on consumerism and the concomitant view of the individual as someone reduced to a consumer. In this view, "The cult of inclusivity" might help "to mask … material differences" (Eagleton, 2016, p. 36) and their social and economic causes. Indeed,

> Nothing is more generously inclusive than the commodity, which in its disdain for distinctions of rank, class, race and gender will nestle up to anyone at all, provided they have the wherewithal to buy it … Everyone is to be included, except those whose politics might undermine the very framework within which such inclusion occurs. (Eagleton, 2016, pp. 32–33)

Inclusiveness should hence be considered as more than a noble gesture of openness. To be *effective*, it requires the creation of a new social consciousness that, through an activist and critical movement, endeavours to *unveil*, and *change*, the political and economic causes that produce exclusion and inequity in the first place.

In this sense, inclusiveness should be a central element in our understanding of mathematical thinking. The problem in front of us is not, I think, to determine whether our mathematical ways of thinking are cultural or not. The problem is to understand their cultural nature. We need to unravel the ways in which the political, economic, and ideological dimensions of society are inevitably entangled in them. To do so, we would need to explore the social problems that contemporary mathematical thinking (inside and outside the school) seeks to address.

Notes

1 By ideology I do not mean something like "false consciousness" or a disguised version of reality. Following Vygotsky, by ideology I mean those elements and standards of a social origin "that have become hardened in the form of legal statutes, moral precepts, artistic tastes, and so on. These standards are permeated through and through with the class structure of society that generated them and serve as the class organization of production" (Vygotsky, 1997b, p. 211).

2 Below, when I discuss Greek theorematic mathematical thinking, I turn to Euclid's *Elements* Proposition 34, where Euclid does not talk about a parallelogram but a "parallelogrammic

area." The Greek and Babylonian different conceptions of shapes are explained by Jens Høyrup in his splendid book, *Lengths, Widths, Surfaces. A Portrait of Old Babylonian Algebra and Its Kin* (see Høyrup, 2002).

3 People from Nawiti took him first for a missionary, then for a government agent. To dissipate misunderstandings, during a Nawiti festivity, after the chief finished his speech, Boas got up and said: "My country is far from yours; much further even than that of the Queen ... My people live far away and would like to know what people in distant lands do, and so I set out" (Boas, as cited in Lévy Zumwalt, 2019, p. 165).

4 The theoretical ordering of the economic as derivative of the symbolic in the anthropological account of society is most intriguing in the works of Marshall Sahlins, who was an ardent opponent of the USA's war in Vietnam and the growing collaborations between the National Academy of Sciences (NAS) and the United States military—a discontent that led him to resign from the NAS in 2013 after more than 20 years of membership. Probably, he saw the problem as primarily symbolic, not economic.

5 The metaphor of the mirror is completely inadequate to deal with our problem at hand: it supposes the mirror and the object as two separate things.

6 The papyrus contains 38 problems stating methods on how to solve problems by means of what the scribe refers to the "the rules (*hoi logoi*) of the geometer" (Asper, 2009, p. 110; see also Fowler, 1999, p. 253). Based on terminological considerations, Asper suggests that the theorematic tradition emerged from the practitioners' tradition. "First, some of the terminology in Euclid betrays a practical origin, for example the term for 'drawing a straight line' (*teinō*, literally 'stretch out'), goes back almost certainly to the aforementioned surveying practices of the 'rope-stretchers.' Similarly, expressions for geometrical entities as angles, certain figures, or the perpendicular go back to craftsmen's traditions" (Asper, 2009, p. 122).

7 The use of pebbles (*psēphoi*) to deal with numbers in Greek culture is very ancient. It is mentioned in Herodotus's (2004) *Histories* (Book II, 36). The Pythagoreans seem to have inspired the use of pebbles before the classical Greek period in the context of philosophical speculation (Lefèvre, 1981). Aristotle (1984, p. 1726, 1092b23–1092b25) refers to Eurytus who, we are told, used pebbles to assign numbers to things: "Eurytus decided what was the number of what (e.g. of man, or of horse), viz. by imitating the figures of living things with pebbles, as some people bring numbers into the forms of triangle and square." The Pythagorean philosophical speculative use of pebbles led to a series of numerical investigations that Robbins calls "arithmology" (Robbins, 1920, 1921) where we find various classification of numbers (e.g., odd and even numbers, polygonal numbers, friendly and prime numbers). These classifications were still discussed at the end of the Hellenic antiquity by the Neo-Platonist (e.g., Nicomachus of Geresa (1924) and Theon of Smirna [see, e.g., Lawlor & Lawlor, 1979; Tarán, 1969]); some still appeared in the curriculum of the Middle Ages (see Boethius's *De Institutione Arithmetica*—English translation in Masi, 1983). Arithmology made its way into Euclid's *Elements* (see Book 2), although deprived of their mystic and speculative origin. Indeed, in Euclid's *Elements*, a part of the Greek arithmology is reformulated *deductively*.

8 Asper (2009) and Netz (2002) are two notable exceptions.

9 When Demosthenes engages in a court dispute against his guardians, he presents a list of his father's estates. The list includes making-knives slaves totalling a value of 19,000 drachmai (see Demosthenes, 2004, p. 20).

10 As Bresson puts it, "From the end of the Archaic to the end of the Classical period Athens was ... the city-state that minted the most coins. Athens enjoyed the advantage of having the silver deposits in Laurion at its disposal. These mines had been in operation in the Mycenaean period ... The date when this coinage began remains uncertain: after 540 BCE at earliest, or perhaps after 530 BCE, is the most likely" (Bresson, 2016, p. 277).

11 Financial documents from the classical period show the use of interest loans. Bankers divided them into two: one pertaining to maritime loans, the other to landed loans. The latter usually had an interest rate of 10% to 12%. In the former, the interest was higher as it presented much more risk: "The interest rate [for maritime loans] was 22.5 percent for a voyage during the season of good weather, and 30 percent for a longer voyage. These were interest rates common at the time. For one-way voyages (*heteroplous*), the interest rate was lower (10–12 percent)" (Bresson, 2016, p. 283).

12 Cohen (1977, p. 69) mentions that, according to records from ca. 329 BC, craftsmen and skilled workers earned around 2 or 2 1/5 drachmai per day.

13 Cohen reminds us that "Employment [of a citizen] by a master for brief periods, a day or two, might have been acceptable to a few [poor] citizens, but extended service was intolerable: in Aristotle's words, 'the nature of the free man prevents his living under the control of another' [Aristotle's *Rhetoric*, 1367a33]" (Cohen, 1997, pp. 70–71). See also Jameson (1974, p. 124).

14 In an introduction to Demosthenes's Speech 36, MacDowell notes that "Pasion [or Pasiōn] had been a slave working in a bank in Piraeus owned by Antisthenes and Archestratus. Eventually he was given his freedom, took over the bank, became a wealthy man, and in return for benefactions to Athens was given Athenian citizenship. In his turn, he owned a slave named Phormion [or Phormiōn] who was a cashier or manager in the bank, was given his freedom as a reward for good service, and later (in 361/0 [BC]) became an Athenian citizen" (MacDowell, in Demosthenes, 2004, p. 150). Over time, Phormiōn became a banker and turned into one of the richest men of Athens. He also ended up marrying his ex-master's wife.

15 Some of those transactions are implied in the stele called *AIUK* 3 (Cambridge) no. 3. This stele shows expenditures and incomes of the Apollo sanctuary on the Island of Delos. Incomes included money from rents and interest on loans to cities and individuals (at a rate of 10%). For the inscription, see https://www.atticinscriptions.com/inscription/AIUK3/3. For an image of the stele see https://www.youtube.com/watch?v=BPpX-piKdPs&feature=youtu.be.

16 Several slab stones discussed in Lang (1957, 1968)—including a marble Salamis table measuring 1.50 m. in length, 0.75 m. in width, and 0.07 m. in thickness—have been suggested as more elaborate examples of the abacus.

CHAPTER 9

Processes of Subjectification

1 Outline

In the theory of objectification, as stated in Chapter 2, learning is conceptualised in terms of two axes: knowledge and knowing on the one hand, and being and becoming on the other. The first axis is investigated through the processes of objectification, while the second axis is investigated through the processes of subjectification. In Chapter 4 we dealt with the processes of objectification. In this chapter we deal with the processes of subjectification. In fact, we have already touched upon them before because the processes of objectification and subjectification go hand in hand. We distinguish them as a matter of methodological convenience. However, so far, the processes of subjectification have been considered only tangentially. In this chapter we look at them in more detail as we now move our focus to the constructs of Being and becoming.

2 The Question of the Subject

To a very large extent, and although rather implicitly, conceptions of the child in mathematics education have been informed by a modern concept of the subject or the individual articulated by philosophers of the 17th and 18th centuries. The subject, as Immanuel Kant pictured it, was seen as moving from the immaturity of previous historical ages to maturity; that is, from being dependent on others in matters of thinking and understanding to being able to use "one's own understanding without the guidance of another" (Kant, 2010, p. 54). To move from immaturity to maturity; that is, for enlightenment to occur, Kant argued that "all what is needed is *freedom*" (p. 55, original emphasis)—the freedom to reason. It is in this context that education was considered for the first time as the space where people's natural capabilities should be nurtured, and freedom exercised: "education should enable people's natural capabilities to find autonomous reason and become personally free" (Jackson, 2007, p. 340). In terms of pedagogical practice Kant advised that education should involve "*guidance* rather than instruction" (p. 340, original emphasis).

The contemporaneity of Kant's enlightened concept of the subject and the ensuing pedagogy appears clearly in the following mathematics classroom example. In a Grade 1 class, Melissa is asked to solve subtraction problems.

 | DOI: 10.1163/9789004459663_010

Melissa's method, the teacher observes, is different from the method that he has in mind. The teacher is afraid to influence Melissa to use his method. After having given several problems to the student, the teacher decides to stop the line of enquiry. He "inferred that if he persisted he might merely train her to behave as he desired rather than encourage her to express her number concept in a novel way" (Cobb, 1988, p. 94). As we see, the question is not about teaching Melissa how to subtract numbers, but rather to make sure that Melissa finds the freedom to think for herself. The question is not about instructing but guiding—and we must stop guiding if our acts risk contravening the student's freedom to reason.

But how did the enlightened philosophers come to conceive of the subject or the individual as they did? The enlightened philosophers sought to build their idea of the modern subject in the overcoming of fear, tradition, and feudal hierarchies. They found, in the concept of *freedom*, the subject's most fundamental trait (Adorno, 2006). The idea of the subject that they envisioned is someone who is not there to follow what others say or do, but one who has to think and reason by him or herself, like in our contemporary Melissa example mentioned above.

Their modern concept of the subject is the result of a lengthy historical and economic process. The emergent craft capitalism that arose in the late Middle Ages and the Renaissance, first in Italy, then in the Netherlands, France, and other parts of Europe (Beaud, 2004), led to new cultural sensibilities and understandings of the self (Taylor, 1989). Such sensibilities broke with tradition and emphasised individualism and the prominence of reason (Horkheimer & Adorno, 2002). They were manifested in the various spheres of human activity. In the arts, for instance, they appeared in the invention of perspective painting, where the world is represented through the laws of mathematical calculation and the subject becomes the origin of what is to be seen, encapsulating thereby "the symbolic relationship between art and worldview" (Wood, 1991, p. 21). The new sensibilities appeared also in the invention of a new painting genre: the portrait (Todorov, 2000), allowing individuals in the Renaissance to express and recognise their cultural experience of the world while asserting the emerging sense of the modern self.

Transposed into education through the influence of Rousseau, Pestalozzi, Piaget, and others, the modern concept of the subject gave rise to a humanist and substantialist concept of the child. The child appeared as a *given entity*; that is, someone already endowed with her own intellectual capacities, who, in order to develop them, simply needs a stimulating social environment (Martin, 2004). A cursory look at the educational panorama of the 20th century shows, indeed, that it is on this concept of the child that the child-centred school of the 20th century, the whole pedagogical movement known as "progressivism," and constructivism were based (Cremin, 1961; Darling & Nordenbo, 2002; Rohrs & Lenhart, 1995). So, in their 1928 book, *The Child-Centered School* (reprinted in 1969), Rugg and Shumaker argued that the school should be built around the child's freedom, interests, and "individual capacity for creative self-expression" (1969, p. 57). The school should offer the child "the ability to govern one's self" (p. 56) and the means to develop to the maximum extent her "wide range of potential abilities" (p. 116). This romantic view of the child is still at the heart of many curricular conceptions. To give but one example, a key document from the Ontario Ministry of Education (2014, p. 3) states that, "All children and students will be inspired to reach their full potential." As we can see, the child is considered to be like a plant that needs light, water, and other environmental stimuli to grow. In this view, the school and the teachers ended up being thought of as fertilisers.

In this chapter, to continue exploring learning, I turn to the child, the student, and the teacher. But I do it from a different perspective. Following the ideas articulated in Chapter 6, I move away from considering the child and the teacher as *given entities*. I draw on a dialectical materialist philosophy where the individual is conceptualised as an entity in flux, in perpetual becoming—an entity who, through practical activity, is continuously inscribing herself in the social world and, in doing so, she is continuously produced and co-producing herself within the limits and possibilities of her culture—much as we saw Plato doing in the previous chapter. I am interested in understanding how, through mathematical ideas and in dealing with others, students and their teachers co-produce themselves and, at the same time, are produced by their cultural-historical context. To understand the dialectical movement between teachers, students, and their cultural context, between agency and affordance, I need to start with a brief discussion of a symbolic structure that, in each culture, defines the *agentic space* of the individuals and provides them with a definite sense of personhood.

3 Semiotic Systems of Cultural Signification

In tune with the results we obtained in the previous chapter, the starting point is that humans are entangled with their social, cultural, and historical contexts. Cultures, indeed, provide their individuals with the raw material of what they are. For instance, the very fabric of human subjectivity in ancient Greece was intertwined with the individuals' participation and their positioning in the economic, social, and cultural activities of the *polis*, such as agriculture, animal husbandry, commerce, religious events, and military campaigns. These economic, social, and cultural activities, out of which a sense of self emerged, provided the individuals with meaning to their lives. It is in this context that individuals in ancient Greece learned to live and die (see, e.g., Vernant, 1993). And so do we in our own cultural-historical context. And because these contexts are different, we find ourselves confronted by a different range of possibilities concerning rights and obligations from those encountered by the ancient Greeks. We find ourselves in front of a world with different political, economic, and legal apparatuses and, as a result, with a different *agentic space*. The scope of the agentic space that shapes the self and agency is both facilitated and constrained by a *symbolic superstructure*.[1] Following the ideas of the previous chapter, this symbolic superstructure arises in the co-determination of the various societal processes, is always in motion, and encapsulates in various tensional and conflictive ways the distinctive features of a culture—e.g., its thematisation of meaning production, conceptions about the relationship between mind and reality, and the understanding of reality itself.

In the theory of objectification, the symbolic superstructure is termed *Semiotic Systems of Cultural Signification* (SSCS). They are dynamic systems that originate in the practical and sensuous activity of the individuals. They comprise interrelated ideas about:

a. the nature of the world (e.g., the nature of mathematical objects and their way of existing),
b. truth (e.g., how truth is and can be established), and
c. the nature of the individuals.

SSCS are full of tensions, as are the activities from where they emanate. They have a (implicit, explicit, or both) normative function and necessarily convey political and ethical views; for example, how we show ourselves to others, and how we are expected to behave socially and be recognised by others.

To understand the operativity of SSCS and how the individual's sense of self is embedded in a web of historical, political, legal, and economic relations, let me mention an example from Athens, in ancient Greece. My example is about two grain merchants, the foreign Artemon and Apollodorus, who borrowed money from the Athenian Androcles. Following a general schema of maritime venture, with the borrowed money, the merchants were supposed to buy goods in Athens, transport the goods by sea to the Pontus (the Black Sea), sell them there at a profit, use the proceeds to buy grains, transport the grains back to Athens, and sell them there at a further profit. From those transactions, the merchants were supposed to pay back the loan and interest and still make some profit for themselves. As was usually the case, the loan required a legal contract to be signed by the money lender and the merchants and some witnesses. This transaction turned out very badly and the loan was not repaid. One of the merchants died, so Androcles took to court the elder brother of the merchants, Lacritus, who had drafted the contract between the lender and the merchants and was in charge of Apollodorus's estate.

There are several points here that define the agentic space and the ensuing sense of self. First, there is an economic web of maritime transactions that has been put into place during the Athenian classic period where money lenders and merchants came into play. Second, there is an Athenian legal system of contracts that serves to regulate the individuals' actions and obligations. Contracts have a definite structure where individuals, places, the purposes and limits of the transaction, the amount of the loan, its interest, the security of the loan, time of repayment, and witnesses, have to be specified. Third, there is a social institution—the court—that the legal system puts in place where the individuals can plead their case in the event of a breach of contract. The legal system recognises rights and obligations under the principle of private property, without which Androcles would not be able to make his case. As he claims in his pleading, "I, men of the jury, lent money to Artemon, the brother of this man Lacritus, in accordance with the mercantile laws for a voyage to the Pontus and back to Athens" (Demosthenes, 2004, p. 134). The nature of the world (the first dimension of the SSCs mentioned above) is shaped by an economic dimension where things are done in certain ways, and by people situated against the backdrop of a differential web of social positions (money lenders, merchants, juries, and so on). It is this dimension that allows Androcles to position himself as he does in the opening address to the jury: "I, men of the jury, lent money to Artemon, the brother of this man Lacritus, in accordance with the mercantile laws for a voyage to the Pontus and back to Athens." Furthermore, at the level of the individuals (the third dimension of the SSCs), there is a social expectation about how individuals should behave vis-à-vis others. In pleading his case, Androcles starts telling the jury that Lacritus and his brothers are not Athenians; they are from Phaselis and, consequently, not trustworthy: "They are the most terrible men for borrowing money in the port, and then, after they get it and draw up a maritime agreement, they immediately forget about written agreements and laws and the need to repay what they received" (p. 133). As we see, Androcles tells the jury that he has acted in accordance with the rules of maritime laws, but the merchants did not. Following rules—subjecting oneself to rules—puts us on a par with others and is, indeed, an important part of the making of the self. In a chapter devoted to kindergarten children I discuss in detail the role of rules and some difficulties that children have conforming to them and acting in the social space that rules create and organise (see Radford, 2020d).

But Androcles still has to convince the jury that his accusation is *true*. How is he to prove it? Here we enter into the second dimension of the SSCS. Of course, it is not enough to *tell* the truth. Truth is always "embedded in structures of authority and assent" (Fernandez-Armesto, 1998, p. 67; see also Shapin, 1995; Detienne, 1996). To tell the truth one is required to enter into a rhetorical structure that Foucault called the society's regime of truth.[2] Androcles has hence to resort to the cultural-historical rhetorical configuration of truth at his disposal, which, in his case, includes a formal procedure recognised by the Athenian legal system: the recourse to witnesses. So, to plead his case, and to *prove* it, Androcles calls upon several individuals to show that what the merchant did was all wrong. For instance, instead of buying grains with the money obtained from selling the goods bought in Athens and the profit made from the sale, the merchants, while being abroad, lent money and bought only a few things. He calls upon Hippias:

> Hippias son of Athenippus of Halicarnassus testifies that ... when the ship was sailing along the coast to Theodosia from Panticapaeum, Apollodorus [one of the merchants] put on board the ship one or two bags of wool, eleven or twelve jars of salt fish, and some goat-skins—two or three bundles—but nothing else." (Demosthenes, 2004, p. 143)

We do not need to continue examining Androcles's case. My point is to emphasise the fact that the *Semiotic Systems of Cultural Signification* operate through a complex web of political, legal, and economic relations. The SSCS come to shape our sense of self by offering each one of us a spectrum of socially recognised positions and allowing us to position ourselves (with the ensuing rights, entitlements, expectations, and obligations) and, at the same time, to be positioned by others. But the most important point is perhaps not to see the self as already *there*, already *given*, moving within a more or less ample agentic space. The self and its space are built and rebuilt constantly out of the ideas conveyed by the SSCS (ideas that are themselves constructed and reconstructed all the time too).

4 Being, Becoming, and Subjectivity

4.1 *Subjectivity*

Pursuant to the ideas discussed in the previous chapter, where we dealt with the concept of culture in the theory of objectification, the relationship between the cultural "raw material" conveyed by the SSCS and the concrete individuals should not be seen in a causal or mechanical sense. On the one hand, as humans, we are unavoidably affected by our cultural-historical concrete context. This is part of our ontological makeup. It is part of what it means to be human. This is the point that the 17th-century philosopher, Baruch Spinoza (1989), made in his *Ethics*—a book that had an important influence on Vygotsky and Marx (Clot, 2015; Fischbach, 2014; Toassa, 2014). However, individuals are not simply affected by their surroundings. They are affected in a *reflexive* manner. What reflexivity means here is that, in addition to being affected by their cultural-historical concrete context, individuals react *agentically* to such context. They do so through the exercise of free will. It is precisely in the scope of the exercise of free will that Vygotsky found the most distinguishing feature of humans within the broad spectrum of natural living beings (del Rio & Alvarez, 1995; see also Tappan, 1998). The discursive and non-discursive practices and technologies that underpin and shape our world do not impede us from thinking and acting differently. There are always possibilities to *interrupt* the quotidian train of our actions and thinking. As Scott puts it, it is in this interruption that "the definitive values that govern thought and everyday action lose their power and authority" (Scott, 1990, p. 4). This interruption makes sense precisely because of our cultural-historical *agentic* nature (i.e., our power to surpass the cultural and historical possibilities on which we draw when we re-act to, and re-enact, the world).

The scope of our agentic nature has been understood in different ways. For those adhering to the idea that we are the only authors of ourselves (i.e., that we are self-making subjects), cultural affordances and constraints are not a matter of concern. Indeed, since the subject is considered to be endowed with a privileged position that makes the subject the locus of meaning and intentionality, the source of knowledge and being—a theoretical conception articulated by the rationalist and empiricist trends of the Enlightenment discussed above—the scope of our agentic nature rests on us and only on us. There is, however, a different concept of the self that considers that, whatever we do and wish, our agentic nature remains confined to "the effect of the subject-positions articulated in discourse" (Atkins, 2005, p. 252). In this view, we are locked inside the discursive universe of our world. Nealon (1998) has cogently argued that this theoretical position rests on an idea of self as *lack*—on a very primal sense, the self lacks wholeness or plenitude.[3]

Historically speaking, feminist, multiculturalist, and dialectical materialist thinkers and scholars in related fields have countered these two conceptions of self and agency by offering a concept of the self as *excess*. In feminist theory, the subject is conceived of as having the power "to reflect on the social discourse and challenge its determinations" (Alcoff, 1988, p. 417). Following a Hegelian thread, for Butler (1999), the subversion of the subject is possible because all acts of signification not only restrict the subject's actions but are, at the same time, in their enactment, always located within the possibility of a variation in the "alternative domains of cultural intelligibility" (p. 185). Dialectical materialist thinkers (e.g., Fischbach, 2014; Macherey, 2008; Mikhailov, 1980; Sève, 2008), have drawn on the third of Marx's (1998) *Theses on Feuerbach* where the subject is featured as one that, while being produced by its circumstances, has, inversely, the power to transform those circumstances. Following Marx, what dialectical materialists add to the agentic conception of the self is that subversion is accomplished in praxis, with others (Fernández Mouján, 2016; Freire, 1998; Guilherme, 2017). Subversion and emancipation appear, hence, not as individual projects but as collective ones (Radford, 2012b, 2013b).

The concept of self that derives from this view and that is at the centre of the theory of objectification is an excess entity that arises out of the *overdetermination* of societal processes: "an emergent property of a historicized experience" (Alcoff, 1988, p. 431); that is, an entity in flux, unachieved and unachievable, a project of life, quotidianly co-inscribing *itself* with *others* into society and culture through concrete praxes.

Thus, what emerges from the effects of affection and the agentive dimension—i.e., the subject—bears the imprint of its culture. At the same time, it always emerges as something different—different from others and from itself: the resulting subject is the difference of the identical (Lévinas, 1978, p. 22); that is an "I" that is not noun but a verb—a verb that does not signify an event (p. 67); an "I" whose formula is "I ≠ I."

This formula captures the conception of the individual as, on the one hand, a dynamic concrete living agentic entity always in flux, in transformation, and, on the other hand, an entity whose agentic dimension can only be understood against the backdrop of culture and history. To refer to the individual in this sense, I shall use the term *subjectivity*. To specify its sense a bit further, I first need to turn to two related terms: *Being* and *becoming*.

4.2 *Being and Becoming*

Being, as I understand it here, is an ontological non-metaphysical category. It is a general, cultural, dynamic (that is, always changing) generative capacity constituted of cultural conceptions of living in the world: ways of conceiving of oneself and of being conceived; ways of positioning oneself and of being positioned. In the previous example from classical Athens, *Being* includes those ways in which slaves, women, merchants, manual labourers, citizens, etc., conceived of themselves and were conceived of by others. Those ways of conceiving of oneself and of being conceived of by others are continuously materialised in the deeds and activities of the individuals. What materialises, however, does not coincide with the capacity that engenders it, for this capacity is a cultural, general, latent capacity. *Being* can only show itself through its materialisations in the concrete world, where it can be recognised as what it is. Being a merchant in classical Athens, for example, is materialised in the deeds of Artemon and Apollodorus, just as being a money lender is materialised in the deeds of Androcles. The always unfolding materialisation or instantiation of *Being* is related to *Being*, but it does not coincide with it. Apollodorus's deeds do not coincide with "merchant" (in the same way as the idea of a triangle does not coincide with any of its materialisations). The *materialisation* of *Being* is what is termed *becoming*.

Now we can come back to the concept of subjectivity. A *subjectivity* is a *unique* sentient, cultural, concrete subject: Androcles, Artemon, Apollodorus, or a student (like Melissa in the example found in the introduction to this chapter) or a teacher. The uniqueness of subjectivity results from the fact that it is continuously *reflectively* affected by *Being* through its concrete materialisations. This reflective affection makes it impossible for subjectivity to slip away or to be replaced (Lévinas, 1978). Subjectivity is always in the making, an entity continuously engaged in a process of be-com-ing: an unfinished and unending project of life.

These processes where *Being* shows itself as always in motion and always different is what we term *processes of subjectification*. They are defined as those processes whereby teachers *and* students *position* themselves, while at the same time are positioned by others against the always contested backdrop of culture and history.

Bernstein (2003, p. 10) has shown that positioning creates "specific relationships within subjects," while Harré (2012) and, more recently, Herbel-Eisenmann, Wagner, Johnson, Suh, and Figueras (2015), have called attention to the fact that positioning involves an ethical dimension. Positioning, Harré suggests, involves a course of discursive actions where "rights and duties are assigned, ascribed, or appropriated and resisted, rejected, or repudiated" (Harré, 2012, p. 196).

To investigate the processes of subjectification and its key idea of co-positionality, we resort to the construct of "voice"—understood as much more than a discursive mechanism: voice as something activity-bound that brings in the postmodern notion of difference and the primacy of the political (Giroux, 1986, 2005). Finding one's voice or having a voice is "moving from silence into speech," "a gesture of defiance that heals, that makes new life and new growth possible" (bell hooks, 2015, p. 29), something that "assumes a primacy in talk, discourse, writing, and action" (p. 33). In investigating the classroom processes of subjectification, we turn hence to the question of the co-positionality of teachers and students through the underpinning constitutive, dynamic, intertwined, social relations and actions, considered through the prism of the critical and ethical stances that they embody.

In the next section I discuss a classroom example.

5 Solving Equations in a Grade 3 Classroom

The following example comes from a Grade 3 class (8–9-year-olds) that we followed as part of a five-year longitudinal research study on the development of algebraic thinking in young children (Radford, 2010, 2018c). In the days preceding our example, the students and the teacher worked on word-problems that the students had to translate into an equation and then solve, using a very simple semiotic system: a semiotic system with three concrete signs: "envelopes," "cards," and the equal sign (=).

The word-problems involve two individuals, each with a number of cards and envelopes. The problem statement specifies the corresponding amounts and the fact that the envelopes all contain a same *unknown* number of cards inside. The problem statement states that both individuals have an equal number of cards.[4]

The word-problems are neither cognitive nor cultural neutral. They are embedded in the first and second dimensions of the SSCS that operates at the level of the school and that, through the practice of what counts as a word-problem, naturalises the manner in which problems are posed and solved. Unavoidably, the problems *show* ostensively some aspects of the *nature of the world,* how it is mathematised, and they provide the basis to illustrate how *truth* can be established. Although the introduction of alphanumeric symbolism was made in Grade 4, in Grade 3, using the aforementioned visual semiotic system, the word-problems and their solutions already conveyed the instrumental and systematic rationality that, historically speaking (Radford, 2006), was a leading feature of the constitution of algebraic thinking in the Renaissance and the early modern period (Viète, 1983; Descartes, 1886).

Figure 44, left, shows the teacher and a three-student group discussing the equation that, in alphanumeric symbolism, corresponds to $2x = x + 5$; Figure 44, right, shows a student during a general discussion solving the equation that, in alphanumeric symbolism, corresponds to $3x + 1 = x + 5$; The "operations" on the "envelopes" and "cards," such as addition, subtraction, and division, remained ostensibly shown by juxtaposing, removing, or grouping signs.

FIGURE 44 Discussing equations written through concrete artefacts

Of course, the class of word-problems that can be "expressed" with such a semiotic system is quite limited, but it is rich enough for students to encounter the basic ideas underlying the simplifications of linear equations to isolate the unknown and find the solution.

The basic ideas of solving equations through algebra rest on two rules: "al-gabr" and "al-muquabala," which are explained by Mohammed Ben Musa Al-Khwuarizmi in his 9th century treatise *Al-Kitāb al-mukhtaṣar fī ḥisāb al-jabr wål-muqābala* (*The Compendious Book on Calculation by Completion and Balancing*; critical edition in Hughes, 1986). The term al-gabr means to complete, to fill, to restore by adding something (a lacking part to a whole). In an equation, say A=B, one restores to A a certain combination of numbers (we would say a certain algebraic expression) by adding to it the quantity of numbers that are missing. So, if A is $10 - 2x$, the expression $2x$ is added to A (and to B), to restore 10. The other rule—that of al-muquabala, which means to confront, to put in opposition—allows us to operate constant and unknown numbers between them, so as to group them in the same expression; in other words, to use a modern expression, to put them on one side of equality (Radford, 1995). In Figure 44, right, we see a student removing a card from the left side of the equation (she will do the same on the right side of the equation). This is an application of the "al-muquabala" rule. With our simple semiotic system of cards and envelopes, the students cannot become acquainted with the rule of al-gabr. The system is too simple. We had to wait until Grade 4 for the students to start using the al-gabr rule. Of course, these rules were not introduced in the classroom through their own historical names and meanings. If I mention them here it is just to remind ourselves of the cultural-historical epistemological thickness of the algebraic ideas that we invite the students to encounter. If I mention those rules here it is also to help ourselves dispel the idea that classroom mathematical knowledge is a-historical and a-cultural. Knowledge, as I argued in Chapter 3, is a cultural-historical entity that is revealed to the students' consciousness through the classroom teaching-learning activity. The *operational* manner in which knowledge is revealed in the school is enmeshed in the assumptions we endorse about how knowledge relates to the world and the rhetorical structures of truth that we make available to the students. And these are precisely two of the dimensions of the SSCS.

Let me now turn to the classroom setting. In accordance with the ideas that I discussed in Chapters 4, 5, and 7, the teaching-learning activity is considered as a collective system in motion. Such a conception of teaching-learning activity allows for the distinction of "moments" of the activity. These moments provide a general dynamic framework for the manners in which the activity unfolds.

In the example that I want to discuss here, the interaction was divided into three moments:

a) *Moment 1:* Students worked in small groups to produce a mathematical text that included: a story of their invention, the translation of the story into an algebraic equation, and the solution of the equation (see Figure 45, Pictures 1 and 2). Each group had a "corresponding" group with which to exchange texts later (see Moment 3 below).

b) *Moment 2*: The group had to send a text to its corresponding group, and vice versa. Each group, working independently, read and evaluated the other group's production (see Figure 45, Picture 3). The evaluation of the other group's production was based on:

1. the *clarity* of the mathematical text (do we *understand* the story, its translation into an equation, and the proposed solution?);
2. the *correctness* of the text (is what the text says *true*?); and
3. the *persuasiveness* of the text (is everything mathematically sound and convincing?).

c) *Moment 3*: Each group met with its corresponding group; at this meeting, each group submitted its comments to the other group and a debate ensued between the groups (see Figure 45, Picture 4; for a more methodological discussion, see Radford and Demers, 2004).

We will limit ourselves here to highlighting passages concerning Moment 3, particularly with regard to the processes of subjectification.

The team of Carl and Sandra—referred to as Team A in what follows—produced the following text:

> For Christmas, Calin got three boxes of Webkinz and Samantha got a box. He [Calin] already has 4 Webkinz. And Samantha already has 28 Webkinz. Now they both have the same amount [of Webkinz].[5]

FIGURE 45 Pictures 1 and 2 show Team A and the corresponding team (Team B) working independently to produce a mathematical text. Picture 3 shows Team B critically examining the other team's text. The other team is doing the same. Picture 4 shows that the members of the two corresponding teams meet to discuss their texts; from left to right, Sara, Sandra, Carl, Elisa, and Christine

The team consisting of Christine, Elisa, and Sara—called Team B—produced the following text:

> Martine has 10 stickers in her collection. She receives an envelope of stickers for her birthday. Cassidy has 6 stickers in her collection. And (receives) 2 sticker envelopes for Christmas. How many stickers are there in each envelope?

The translation into equations and the equation solutions are shown in Figure 46.

When Moments 1 and 2 were concluded, the teacher invited the groups to meet; she reminded the groups of the points to focus on in the text (i.e., clarity, correctness, persuasiveness). After she organised the groups, she left to see other teams.

Team A begins the discussion:

1. Carl: Um, what we liked was that your story was clear, there were no mistakes, we could read it well. That's what we liked about your story.
2. Sandra: (*Indexically pointing out the other team's solution approach*) It's good, like doing it [like that] ...
3. Christine: Is that what you liked about the equation?

4. Sandra: And here, here (*Pointing to the solution of the equation*) what we liked is that you put the envelope equals 4.

Team B continues the discussion by giving the reason for its judgments:

5. Elisa: We really liked your story, because I love Webkinz, but the only thing is … you didn't ask … it was …
6. Carl: (*Interrupting*) No, that's … you say what you like.
7. Elisa: Oh okay. What I like … about your story is that … I can see it's very clear, no smearing; you can even see everything … I saw that, you even wrote the words to [say] what. (*She points to the drawings.*) You showed what the boxes are, like here … (*She points to the words "boxes" in Carl's group equation.*) … and what I liked about your solution is that, I saw that you, you circled well … you didn't smear at the top, yeah. (*They laugh.*)

We can see in this short passage how the two groups begin their discussion. They read, interpreted, and made a first judgement on the mathematical text of the other group. The students begin by highlighting the elements that they find well done. This approach, which may seem to be of little interest, allows, however, the students to enter into mathematical practice and to start positioning themselves within it. In the absence of the teacher, they organise themselves. On line 5, Elisa begins a comment on the inadequacies of the other group's text; on line 6, Carl reminds her that this is not the time yet. Certainly, the mathematical language is still fragile; the fluid use of technical terms in mathematical discourse still eludes the students. Thus, in line 7, Elisa says: "I saw that you have, you have circled well …"; the sentence remains unfinished. But the other group understands that she is referring to the objects removed in the simplification of the equation (i.e., the application of the "al-muquabala" rule; see Figure 46, bottom right). This is part of the process of subjectivation through which students encounter other voices, come to make room for themselves, and take a perspective in the social world. Equally important in the unfolding collective process of subjectification is the fact that the students come to demonstrate *empathy* for others: they strive to *understand* the text of the others.

Then, the groups move to the discussion of what they thought could be improved in the other group's text.

8. Carl: In your equation, you didn't put the equal sign (*See Figure 46, top*).
9. Sandra: And you have to put that in!
10. Carl: You always have to put that. This is what you should have done. (*Points to their own page and shows them.*)
11. Christine: It was Elisa and Sara's idea!

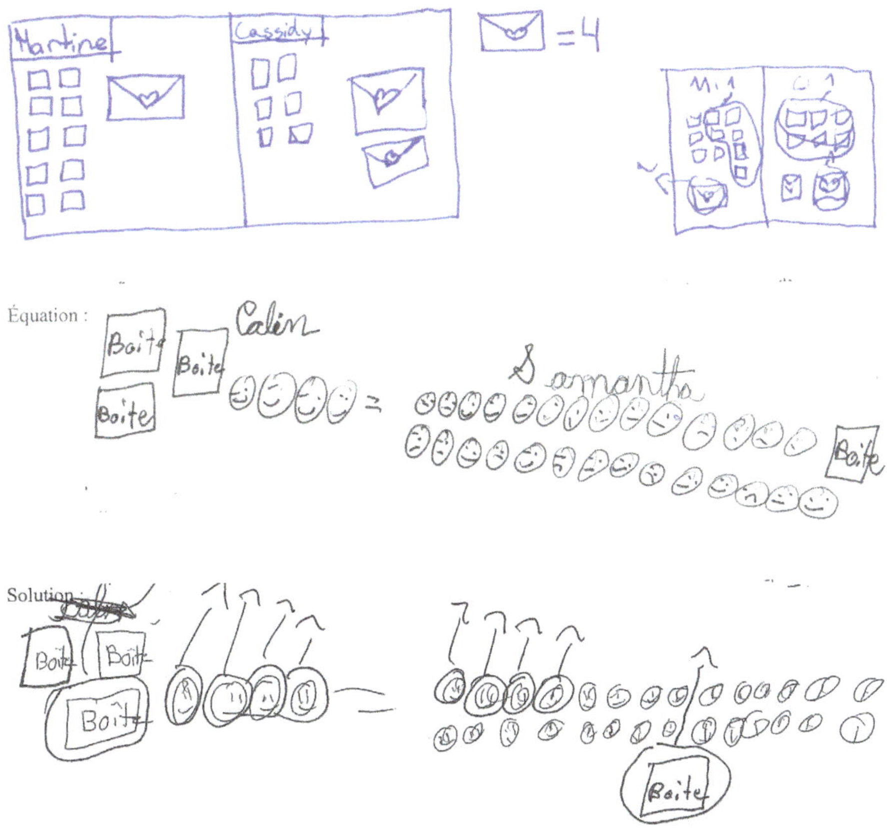

FIGURE 46 Top row: equation and solution of Team B. Bottom row: equation and solution of Team A

12. Carl: (*Referring to the simplification of the equation*) You didn't do one by one. Also (*Referring to the solution of the equation*) 4 what? We don't know 4 what? You just wrote 4; you should have written "cards."
13. Elisa: … not cards, "stickers"; they are "stickers."

As we have just seen, the students of Team A point out three elements:

a. The absence of the equal sign in the equation of Team B (lines 8, 9, and 10).
b. The procedure for simplifying the equation (the "al-muquabala" rule): instead of removing the objects on each side of the equation one by one, the students removed 6 objects *at the same time* on each side of the equation (line 12).
c. Lack of return to the original problem to identify the nature of the solution ("4" versus "4 stickers") (lines 12 and 13).

The students of Team B accepted a) and c), but did not accept b). The teams' dispute concerned the rhetorical structures of truth.

Then, it was Team B's turn to offer its criticism. Team B's criticism focused on the lack of questions in Team A's text. The criticism led to intense debate. The following is a short excerpt from the debate:

14. Elisa: You didn't say "how many ..." like, "how many are in each box?"
15. Sandra: You are not supposed to say how many ... in each box!
16. Carl: (*In agreement with his teammate*) You're not supposed to!
17. Sandra: (*Explaining further*) 'Cause you're supposed to find out how much is in each box!
18. Elisa: (*In agreement with her teammates*) Yes, you have to ask like, let [see] ... like [we] wrote ... "How many stickers are in each envelope?" That's what you need to look for!

Elisa has difficulty articulating her criticism. The conversation does not seem to be about the same object of discourse. Carl and Sandra think that Elisa is asking them to mention the answer in the story they have composed. The teams were not able to reach an agreement. They decided to continue the discussion on other points. The discussion turned to the solution of the equation. Team B complained that the objects have been drawn too small, but the students eventually agreed that the size of the drawn objects does not matter. Size can be ignored. It is part of the abstractedness of mathematical thinking. They also discussed the kind of numbers used in Team A's text. Team B suggested that it takes too long to solve an equation with a large number such as 28. The students of Team B felt that if Team A had used 100, it would have taken about 10 minutes to solve the equation. Group A suggested using circles for large numbers, thus creating a new symbolic notation. The discussion was not taken further, but the idea remained on the table. And, in fact, this is the route this class will explore next year with another teacher.

The previous passages show the processes of objectification and subjectification. They show, indeed, the students' encounter with cultural ways in which problems in algebra are formulated, translated into an equation, and solved, while showing at the same time how the students position themselves and are positioned by others. There were several points that remained unsettled (e.g., the correctness of removing several items at the same time; whether a question has to be included in the text), and, to go further, the teacher's explicit intervention was required.[6]

When the students were wrapping up the discussion on large numbers, the teacher arrived to see the group.

19. Teacher: What things did you notice that could have been improved in the other group? Okay. We'll start here. (*She points to Sandra.*)
20. Sandra: We saw that they didn't put the equal sign.
21. Teacher: Ah! So, the equal sign is missing.
22. Sandra: And then, here, they didn't put it on one by one, they just took one group and they took it off (*Referring to the "al-muquabala" rule*).
23. Teacher: Okay. I agree with [the remark about] the [missing] equal sign because that's an equation, right? So, that would be one thing to improve next time? Add an equal sign.
24. Christine: Yes …
25. Teacher: The idea that they circled all 6 [cards], and then just pulled out all those cards at once, I don't have a problem with that.

Next, the students reported back to the teacher the critique from Team B concerning the missing question in the Team A text, because, as Christine said, without a question, "you don't know what you [should] do!"

26. Teacher: (*Talking to the students of Team A*) What do we want to know?
27. Carl: … how much is in the box.
28. Teacher: Oh! We want to know how much is in the box. Do you think that in a situation like this, for someone who reads this [the text], and wants to … and then wants to find the solution … do you think it's important to add the question? (*See Figure 47, top.*)
29. Sandra: Yes, I do …
30. Carl: I say no, because in a story, the adventures, we don't say like … what's the solution! (*See Figure 47, bottom.*)
31. Teacher: I think that in a story like this, it's important to have a question, if you want to know how many Webkinz there are in each box.

In the end, there is no consensus. Carl saw the mathematical text a bit like a literary text, where you find a situation; then the situation or "adventure," as he himself calls it, unfolds, which would correspond to the simplification of the equation; then followed by the end, which consists of naming the solution.

In a sense, the teacher's intervention could be seen as a failure to institutionalise knowledge. Seen from this angle, in the preceding passages, the teacher would have missed the opportunity to indicate to the students how a problem is posed and in particular, the role of the question. On closer inspection, the teacher certainly took a stand: (see lines 21, 23 and 25 above). But she tries to do so without imposing her authority. The issue is not to get rid of authority, but

FIGURE 47 Top: the teacher offering a different interpretation concerning the place of the mathematical question. Bottom: Carl refuting the teacher's interpretation

to exercise it responsibly; that is, by taking into account the voices, gestures, and perspectives of others. In the theory of objectification, the teacher is considered the personification in the classroom of a cultural rationality; but her goal is not to impose this rationality—which would be the case of a pedagogy of reproduction. In the theory of objectification, the teacher's goal is rather to *offer* the students another way of looking at the problem (a historically and culturally constituted way of thinking about mathematical problems) for students to consider it from their own perspective.

A process of objectification does not consist of imposing a way of looking at the world and subjecting the student to it. Objectification consists of an always unfinished, *critical* encounter with knowledge. And it is precisely this offer that the teacher makes to the students in line 28: "Do you think that …?" She enters into a *dialogical* situation that necessarily makes her *vulnerable*. She can be refuted. But to be refuted is not to be avoided. On the contrary: to be refuted is part of the collective co-positioning of voices and subjectivities—those of the students but also those of the teacher. In this line of thought, where the teacher does not play the role of a patriarch of knowledge, the teacher is not above the students. She is *with* the students; she works *with* the students. This is the meaning of the concept of joint labour in the theory of objectification.

As a matter of fact, as previously seen, the teacher is refuted in line 30. So, at line 31, she takes a position, but, again, in harmony with the ethical project of teaching-learning, without imposing her authority. The educational project revolves around the creation of conditions that allow students and teachers to participate in classroom activity based on non-alienating forms of human interaction and collaboration, and through which teachers and students express and position themselves in the public space and, in so doing, co-produce themselves daily. From this viewpoint, at line 31, the teacher can go talk to other teams. She has already reached her didactical goal. Of course, the discussion will continue, and new contradictions will be revealed during the general classroom discussion and the following days and years.

6 Synthesis

I started this chapter by putting into question a concept of the student that has had an important influence in recent understandings of teaching and learning—a concept that lies at the heart of child-centred education. I attempted to show that this concept of the student is a transposition to the educational field of a more general notion that was articulated by philosophers of the Enlightenment and that came down to us through the influence of Rousseau, Pestalozzi, Piaget, and others. In this line of thought, the student is conceptualised in a humanist and substantialist manner: the student is perceived as a *given entity*—i.e., an entity already endowed with her own intellectual capacities. To develop these capacities to their full extent, the student simply needs an encouraging social environment. In contrast to this view, which reduces the social and the cultural realms to a mere stock of stimuli, and following the ideas of the previous chapter, I suggested a view of the students according to which the student is always in the process of becoming. And so are teachers. In this view, the becoming of individuals in general, and teachers and students in particular, is entangled with their social, cultural, and historical contexts. It is in culture—and more specifically, in the Semiotic Systems of Cultural Signification (SSCS)—that individuals find the raw material of their subjectivity. The SSCS is a dynamic symbolic superstructure that offers cultural conceptions about the nature of the world and the nature of the individuals. This symbolic superstructure shapes the individuals' agentic space that is critical to the ensuing sense of the self. To illustrate those ideas, I discussed an example from ancient Athens that gave us the opportunity to see Androcles as both producing himself through his deeds and, at the same time, being produced by his cultural-historical context. Thus, to position himself in the Athenian social practices, Androcles draws on the agentic space at his disposal; he also draws on his society's regime of truth. Then, to formalise those ideas and to better grasp them theoretically, I introduced the concepts of Being, becoming, and subjectivity. Against the background of the cultural-historical category of Being, subjectivity is considered an entity in perpetual becoming: the always partial and unachievable *product* of collective processes of subjectification: neither an epistemological subject, nor a cognitive one, a subjectivity is, in this view, the concrete subject that breathes, transpires, dreams, suffers, and hopes with others.

In the second part of this chapter I tried to show, through excerpts from a Grade 3 class example, some elements of the processes of subjectification on which the teacher and the students embarked as they engaged in a mathematical teaching-learning activity. The example also shows a process of objectification. The simultaneous occurrence of those processes results from the fact that they are intertwined with each other. The reason for their intertwining is found in the very nature of mathematical learning, as was defined in Chapter 2: in a process of subjectification—in trying to position themselves in a mathematical practice—the students resort to mathematical *knowledge*. And it is here that the processes of *objectification* come into play. Reciprocally, by engaging in a process of objectification—in trying to encounter cultural knowledge—the students are led to *position* themselves vis-à-vis knowledge, and this positioning is part of the processes of *subjectification*. This intertwining of processes of objectification and subjectification is but a practical expression of the fact that learning is unavoidably about knowing and becoming.

The classroom example shows, in particular, the way in which the students encountered some aspects of algebraic knowledge, such as knowing how to pose a problem, translate it into an equation, and solve the equation algebraically. They engaged in intense discussions through which they encountered a historical-cultural knowledge that the school as a social institution places in a privileged position. Far from neutral, the privileged position of such knowledge is a response to societal needs that are defined through a complex process that is ideological, political, and economic. My intention here is not to enter into the politics of the curriculum (for discussions on this topic see, e.g., Baldino & Cabral, 1998; Pais & Valero, 2012; Popkewitz, 2004, 2009; Valero, 2004; Valero & Knijnik, 2015). However, I cannot fail to mention that the body of knowledge that any school brings to the sphere of teaching and learning embodies tensions and interests of diverse societal groups. In the case of the lesson I discussed here, mathematics appears as a necessity: mathematics is expected to equip the students with a powerful conceptual tool to enter the job market and arm them with the rational, instrumental, abstract, and calculation-oriented skills to engage in the economic consumerist activities required by the neoliberal market. What the students are learning is hence not just *mathematics*. What they are learning is to think along the lines of a definite form of cultural rationality whose politics and excesses will need to be addressed (Radford, 2018c, 2020e).

Coming back to the classroom examples, they show that the encountering of cultural knowledge targeted by the curriculum was accomplished through a complex organisation of interaction. Such organisation seeks to enable the students to meet other voices, to take a critical stand in the discussions and thus co-position themselves in the classroom mathematical practice.

Naturally, learning is an unfinished process. There will always be a possibility to go further, both in terms of knowing and becoming. In our example, the algebraic unknown was represented by the number of objects in a container (an envelope). The number of objects in each container must always be the same (otherwise the lesson would end up dealing with several unknowns at the same time). This point, however, was not the subject of theoretical thematisation during the discussion. The teacher will have to come back to this point later on in order to raise awareness of this essential element in dealing with equations. Also, the teacher and the students will need to come back to the question of *responsibility*. We have seen that, when Team A presents its criticism about the lack of an equal sign in Team B's equation, Christine, on line 11, tries to relieve herself of responsibility vis-à-vis the critique; she says: "It was Elisa and Sara's idea!" The students have to realise that they do not belong to a team only when things are going well!

At any rate, the classroom examples show a complex didactical organisation that sought to offer conditions of possibility for the students and the teacher to collectively produce knowledge and to engage in human forms of collaboration beyond the self-centred forms of individualist and transmissive learning. It is within this complex didactical organisation that an agentic space was forged and that the students and the teacher co-positioned themselves in an unfolding and open-ended process of subjectification. In this process, the teacher and the students accepted to leave their own safe niche to go towards the unpredictable. The Hegelian philosopher John Russon sees an artistic element in this act through which we throw ourselves beyond the limits of our daily comfort. Artistic, because this act is an act of "homemaking"; that is, through this act we forge our place in the world: "It is only through … risking ourselves—exposing ourselves, beyond the comfortable terms of familiar life, to an unknown, beckoning alien reality—that we *grow*, that we come to inhabit a deeper, richer, and more substantial home" (Russon, 2017, p. 39).

Acknowledgement

Part of this chapter draws on Radford, L. (2018). Semiosis and subjectification: The classroom constitution of mathematical subjects. In N. Presmeg, L. Radford, M. Roth, & G. Kadunz (Eds.), *Signs of signification. Semiotics in mathematics education research* (pp. 21–35). Cham, Switzerland: Springer. Used with permission of the copyright holder.

Notes

1 Ratner makes an important disctinction between individualistic and cultural conceptions of agency. In the former, the individuals are considered to construct their own significance of things and choices: individualistic agency "explains facts about social phenomena in terms of facts about individuals" (Ratner, 2000, p. 413). In the latter, "agency always operates within and through a social structure" (p. 421).

2 Foucault noted that "Each society has its regime of truth, its 'general politics' of truth: that is, the types of discourse which it accepts and makes function as true; the mechanisms and instances which enable one to distinguish true and false statements, the means by which each is sanctioned; the techniques and procedures accorded value in the acquisition of truth; the status of those who are charged with saying what counts as true" (Foucault, 1980, p. 131).

3 A few years ago, a colleague and I were visiting a university in Denmark on the occasion of a thesis defense. During the evening, returning from the university, we made a stop at the hotel bar. Inspired by the thesis defended that day, the conversation turned to the question of agency. To counter my argument against the defeatist idea that we are inescapably bound by the discursive and non-discursive practices and surveilling mechanisms of our system (the capitalist system), my colleague explained that my argument was indeed a production of the same system it was trying to get away from; my argument was produced by a kind of "safety valve" built into the system, and that there was even a technical name for my argument: *phantasy*.

4 A few years ago, this type of problem was experimented with in classes of older students; see Radford and Grenier (1996). For a more recent account, see Janßen and Radford (2015).

5 Webkinz are small stuffed animals.

6 I say "the explicit *intervention* of the teacher"; not the "explicit *presence* of teacher," because the teacher has been present all the way along. Her presence transpires in the social organisation of the classroom as well as in the mathematical task in which the students are engaged.

CHAPTER 10

Ethics

> What makes us unique is not our personal qualities (the ways we can rise above other people's definitions of us) but precisely the qualities of our subjections ... It is outside the self that we need to look for the conditions of agency, responsibility, and ethical subjectivity.
>
> NEALON (1997, pp. 145–146)

∵

1 Outline

The first chapters of this book offered an overview of the theory of objectification and its place in the landscape of contemporary socio-cultural theories. As we moved from chapter to chapter, the concepts introduced at the beginning—concepts that appeared first in a *general* manner—were discussed and presented in more detail. In the terminology of dialectical materialism, our journey has been an ascent from the abstract to the concrete. In the course of this ascent the various concepts acquired progressively specific determinations showing, as we proceeded, a more definite content as well as the concepts' interconnections and interwoven meanings. In this final chapter I discuss something that has always been there, since the very beginning of Chapter 1, but that emerged bit by bit in previous chapters and that comes to full light at the end: the question of ethics.

In the first section of this chapter I contend that any teaching-learning approach or model rests on a certain ethics. In other words, ethics is an omnipresent element of mathematics teaching and learning. By way of example, I discuss the ethics that underpins direct teaching and constructivism. Then, to pave the way for a discussion of ethics in the theory of objectification, I explore summarily three ethical systems: those of Kant, Hobbes, and Lévinas. I try to show that these chief systems of Western ethical thought fail to take into account the constitutive cultural-historical-political mediating relations that pervade our relationships with others. I argue that an ethics oriented towards a pluralistic sense of collectivity, inclusiveness, and the common good cannot be conceived of as a mere encounter of Self and Other. It should rather be

 | DOI: 10.1163/9789004459663_011

considered as a *liberating* force that can help us come to grips with the historical, political, and economic sources and structures of oppression, violence, and inequality. This ethics—conceived of as an *ethics of liberation*—should be at the centre of education. However, the conception of the contemporary school as an appendix to political economy narrows the scope of ethics to processes of an instrumental and utilitarian nature. We are in the grips of a "fatalistic ideology which, taking a despotic approach to education, reduces it to mere training in the employment of technical dexterity or scientific knowing" (Freire, 2004, p. 102). Unfortunately, our teaching-learning approaches are turned towards the practice of alienation rather than to the practice of liberation. The rest of the chapter is a sketch of the ethics of the theory of objectification—what we call a *communitarian ethics* that, emphasising the collective joint labour of teachers and students, seeks to offer reference points towards the renewal and imagination of new forms of relating to others and working with others.

2 The Ineludible Presence of Ethics in the Mathematics Classroom

Let me start by suggesting that classroom teaching-learning activities, regardless of the underpinning pedagogical models, are based on two interrelated axes:

1. the axis of the *forms of human collaboration*, and
2. the axis of the *forms of knowledge production*.[1]

The second axis refers to how *knowledge* is *produced* and *circulates* in the classroom. It includes assumptions about the nature of knowledge and how truths about knowledge are validated and refuted (e.g., forms of inquiry and proving). It also includes questions about *who* produces knowledge.

The first axis includes kinds of interaction with others, the manners by which we relate to others, and brings to the fore questions of responsibility, power, and autonomy.[2]

This basic, underlying composition of teaching and learning activities helps us to understand why any pedagogical model and any approach to teaching and learning rests on a certain ethics.

First, learning, by involving *interaction* between people (the first axis), be it synchronically or diachronically, in face-to-face or virtual (digital) formats, is unavoidably based on relations with others (relations of power and subjection; relations of solidarity and emancipation, etc.).

Second, but no less important, learning always involves conflicting views about what counts and matters in learning; it also involves the *legitimation* of particular forms of knowledge and knowing.

Ethics, as understood as that which frames and orients our relationships with others, is present in these two points. Perhaps what is surprising in the claim that I am making is not that ethics is omnipresent in teaching and learning but the fact that, some exceptions put aside (e.g., Atweh, 2014; Atweh & Brady, 2009; Boylan, 2016; Dubbs, 2020; Ernest, 2009; Maheux & Proulx, 2017; Maheux & Roth, 2014; Roth, 2013; Silva D'Ambrosio & Espasandin Lopes, 2015; Walshaw, 2013), we have not been able to notice this presence before—or at least not with the importance it deserves. The roots of that oblivion or omission are entrenched, I suggest, in the long-standing conception of education as a technical act (most crudely visible in the case of mathematics education), a conception that reduces education to a matter of Knowledge—acquiring Knowledge—overshadowing with this same movement the dimension of Being. But once we overcome this restrictive vision of education (mathematical, scientific, artistic, etc.), ethics not only appears as an element always present in teaching and learning but becomes something essential to take into account. For one thing, the ethical nature of classroom interaction and knowledge legitimation are dialectically associated with how teachers and students understand their engagement and responsibilities in the mathematics classroom and see themselves as practicians of school mathematics.

Let's look at direct teaching and try to find its underpinning ethics. The teacher chooses some problems and shows the students how to solve them. Then the students repeat the problem-solving procedures shown by the teacher. Choosing the problems, showing how to solve them, imitating the teacher's problem-solving procedure, and having eventually a classroom discussion are part of the forms of classroom knowledge production and circulation. What are direct teaching's forms of human collaboration? They are configured around the recognition of the teacher as the authority and the subjection of the students to the teacher's authority. What we find in this pedagogical model is the *ethics of obedience*, the submission of the student to the teacher, and the power and authority of the teacher over the student. This ethics comes to modulate the social relations we see in direct teaching classrooms. If we now turn our gaze to constructivism, what we find is the *ethics of the student's autonomy*, of her freedom, of her right to construct her own knowledge. It is this ethics that modulates and attunes the social relations between the Grade 1 student Melissa and her teacher discussed in Chapter 9. The reader may recall that the teacher refrained from asking Melissa questions about how to subtract numbers. By stopping his questions, the teacher is following the constructivist ethics that orients his relationship with the student and frames his understanding of the unfolding situation.[3]

What is at stake in mathematics teaching-learning from an ethical perspective is the question of the nature of human relations and conduct. Of course, this is a problem that has kept philosophers busy since Aristotle. One of the questions that philosophers have discussed at length, and in particular recently (Bauman, 1993; Critchney, 1999; MacIntyre, 2007), is whether human relations and conduct can/should be framed in terms of general principles about our duties and obligations; that is, whether ethics can be understood as *deontology*. Within the framework of postmodernism and its skepticism about grand narratives, ethics has come to be seen as more contextual and subjective.

Before offering an overview of ethics as conceived in the theory of objectification, in the next sections I would like to turn to three ethical approaches. I shall briefly deal with ethics in Kant, Hobbes, and Lévinas.

3 Kant

Kant was able to formulate a deontology of human action. But in order to do so, he had to make assumptions that came with a high price. Indeed, first, he had to assume that individuals are by nature provided with the same reason—a legislative Reason. Second, the individuals' natural reason moves in a world of laws that can define the social order. This is why Kant does not need to base his ethics on something external to the subject, such as God's will or the revelations of God. From those assumptions, Kant can formulate his famous *categorical imperative*. The categorical imperative is not an instrumental principle; that is, something that we can use to reach an end. On the contrary, it is a principle that is stated in the first person. It refers to an action independent of the purpose that the particular action might have. The categorical imperative answers the question: *What should I do?* and sets out the concrete action of a subject in terms of possible actions that could be taken by *any* individual. Kant enunciates the categorical imperative as follows: "*act only in accordance with that maxim through which you can at the same time will that it become a universal law*" (Kant, 2006, p. 31, original emphasis). As Arendt explains,

> The point of the matter is very simple. In Kant's own words: I can will a particular lie, but I "can by no means will that lying should be the universal law. For with such a law there would be no promises at all." Or: I can want to steal, but I cannot will stealing to be a universal law; because, with such a law, there would be no property. (Arendt, 1992, p. 17)

The categorical imperative is a *singular* made *universal*; it encapsulates the individualism of its time. MacIntyre (1966, p. 209) says: "The eighteenth-century individualist sees the good as the expression of his feelings or the mandate of his individual reason because he writes as it were from outside the social framework. Society appears to him an aggregate of individuals." The categorical imperative is an attempt "by the individual to supply his own morality, and at one and the same time, to claim for it a genuine universality" (p. 208). Indeed, the categorical imperative makes otherness appear as a *reflection* of the self in a mirror through which the self can see how the action of the other might affect *itself*. One of its problems is that it positions the action of the subject in such a way that it exempts it from any questioning regarding possible forms of violence that such action may produce in the marginalisation of social groups and their oppression.

The formulation of morality in terms of universal laws allows Kant to try to solve two problems at once: the problem of power and the problem of freedom.

With respect to power, what power do I have to enunciate the categorical imperative? How do I not impose my world view on others through this imperative? The authority of my imperative can only be justified on the assumption that it does not come from my own choice but from the way in which that imperative suits each individual, since each individual shares with me the same reason.

With regard to freedom, to what extent can we claim to be free if our actions are subject to categorical imperatives? For Kant, the freedom and emancipation of the individual resides precisely in her use of that which brings her out of the darkness of feudalism: her use of reason; not just any reason but the reason that Kant considered to be universal, and which dignifies the individual and elevates her from the obscurantist world and the whims of childhood to culture (see Chapter 9). It is the use of this reason that embraces the categorical imperative that makes the individual will free from any inclination and good in itself. "It places individuals as freely acting agents bounded by universal, global values" (Popkewitz, 2009, p. 252). Thus, subjection to the moral law appears in Kant as the affirmation of rationality, of our freedom and autonomy. As Nodari notes, in Kant,

> for the will to be autonomous, the force of autonomy must not come from an external source alien to the subject herself, but only from reason itself. In other words, the autonomy of the will means only the capacity of the human being to give herself the moral law as a universal value. Only a universal legislating will, that is, an autonomous will, is capable, at the same time, of giving itself the law, and also of understanding itself, and of recognising itself as obedient to the law, because only the rational

> being is capable of self-legislation; that is, capable of being autonomous. (Nodari, 2010, pp. 61–62)

By assuming a universal reason, reasonable in itself, Kant can solve the problem of the tension between the individual and the social world and the tension of human action in that world. However, the theoretical weight of Kant's assumption is so great that it ends up sinking his ethical project. Yet, Kant's impossible assumption allows him to solve the problem of Thomas Hobbes's contractual ethics articulated a century earlier—an ethics that revolves around the aspirations and inclinations of the individuals.

4 Hobbes

In Hobbes, in fact, ethics revolves around the range of fundamental human motives. In *Philosophical Rudiments Concerning Government and Society*, such fundamental human motives are ultimately reduced to the desire of the individuals to dominate and "through desire of preserving themselves and by mutual fear" (Hobbes, 1841, p. 70). Ethics in Hobbes appears as an *agreement* or *contract* that sanctions behaviour between subjects to avoid aggression against each other. Hobbes says:

> the dispositions of men [sic] are naturally such, that except they be restrained through fear of some coercive power, every man will distrust and dread each other; and as by natural right he may, so by necessity he will be forced to make use of the strength he hath, toward the preservation of himself. (Hobbes, 1841, pp. xiv–xv)

He goes on to say that

> the state of men [sic] without civil society, which state we may properly call the state of nature, is nothing else but a mere war of all against all; and in that war all men have equal right unto all things … all men as soon as they arrive to understanding of this hateful condition, do desire, even nature itself compelling them, to be freed from this misery. (Hobbes, 1841, p. xvii)

With Hobbes we arrive at a conception of civil society as a society made up of monadic individuals formally bound by contracts that guarantee to counteract natural dispositions. In his commentary on Hobbes's ethics, MacIntyre points out that arrangements or covenants that seek to ensure security between individuals lead to a social contract that comes to regulate the forces and powers of each: "To give the covenants the backing of a sword there has to be an initial contract by which men transfer their power to a common power which becomes sovereign among them" (MacIntyre, 1966, p. 133). These "contracts," which permeate all spheres of social life, are not necessarily explicit. They do, however, bear the hallmark of each other's expectations for maintaining social order, and they try to leave room to fulfill, as much as possible, personal aspirations and to safeguard each other's security. Historically speaking, it is Brousseau (1978, 1988) in the field of mathematics education, it seems to me, who was the first to have noticed the ubiquitous presence of those contracts in mathematics classrooms. They appear as didactic contracts, as Brousseau rightly calls them. They reflect an underpinning conception of the classroom as a contractual ensemble of Hobbesian individuals.

5 Lévinas's Ethics

5.1 *Ethics as Responsibility*

Lévinas starts from the recognition that all Western ethical systems are configured around the idea of the self.[4] In those systems the self is both a constituted consciousness and a constituting entity, which, already established, enters into a relationship with others. In this view, ethics is a term connecting two poles—self and Other—in more or less convenient ways (as in Kant and Hobbes). More importantly, the very fabric of the connecting term is knowledge. For example, without recourse to knowledge, Kant's categorical imperative cannot be stated. The same goes for the contractual term that links self and Other in Hobbes's system. Lévinas takes a different route.

First, he removes the self from the privileged seat with which it has been traditionally endowed and from where it has conceived of itself as a constituting consciousness and the principle of the ethical relationship. In Lévinas's ethical account, it is not from the self and its deeds that ethics appears. Stripped of its imperialism, the self appears conceived of as the *result* of an ethical relationship.

Second, the very fabric of the ethical relationship is not knowledge; it is a *presence*—the Other's presence that comes into existence not through the cogitations of the mind. The Other's presence comes to us in a sentient manner: through the proximity of the *body*, through *touching*—the physical skin-to-skin touch or touching from a distance with the eyes.

To understand Lévinas's ethics we have to bear in mind that, for him, the original experience of the self is the human encounter and the idea that human beings find the greater meaning of life in their relationships with other human beings (Nodari, 2010). Rather than a threat, rather than being opposed to us, the Other is our *liberation*. The Other is the only possibility that we have to overcome our narcissist tendencies and our solipsist niche. Because Lévinas's ethics starts from a de-centred ego, his ethics is not one of reciprocity. It is an *asymmetrical* ethics where we live in the constant refusal to *assimilate* the Other. "What counts here," Lévinas says, "is the refusal to tame or domesticate [the Other] through a theme" (Lévinas, 1978, p. 158); that is, through the imposition of my own conceptual categories and perceptions. The ethics that he proposes is an ethics of generosity in which the Other, with its simple presence, offers us the extraordinary gift of being able to give. Being is *being-for*: an entity turned towards the Other in a heterogeneous relationship that, in the social area of daily life, manifests itself as an *opening*. In this context, "the presence of the Other is the very meaning of our existence" (Noradi, 2010, p. 172).

Lévinas does not understand our subjectivity as something that emanates from the self. Subjectivity is exteriority. For him, subjectivity is based on hospitality, on welcoming, and hosting the Other. "Truth, now, is no longer the adaptation of the object to the consciousness of the thinking self but is the receiving of the other" (Nodari, 2010, p. 177). In encountering and receiving the Other, we remove our armour and appear in a "nudity" that exposes ourselves to effusion (Lévinas, 1978, p. 116). We expose ourselves and appear in our vulnerability. In encountering the Other we embrace it with disinterested responsibility.

In Lévinas's view, ethics is not an ethics *of* responsibility. Ethics is understood *as* responsibility. In his book *Éthique et infini*, Lévinas (1982) notes that responsibility is "the essential, primary and fundamental structure of subjectivity ... It is in ethics, understood as responsibility, that the very node of the subjective is knotted" (p. 91).

He goes on to say that "[r]esponsibility in fact is not a simple attribute of subjectivity, as if the latter already existed in itself, before the ethical relationship. Subjectivity is not for itself; it is, once again, initially for another" (Lévinas, 1982, pp. 92–93).

5.2 *The Flesh*

Lévinas's ethics rests centrally on the idea of the flesh. But in opposition to other philosophical systems in which the flesh and corporeality remain thematised within the confines of cognition and consciousness, Lévinas offers an account where the flesh and the individual's corporality more generally are active parts of our encounter with the Other.

Lévinas proceeds to an original phenomenological analysis of the body. The body is neither the origin nor the result of consciousness.He departs here from both empiricism and idealism. In *Autrement qu'être* (Lévinas, 1978), he insists that "an idea is not a simple sublimation of the sensible" (p. 102). The body appears as a body-sensibility posited, as it were, in a *limit position*. Indeed, the materiality of the body drives the self towards a radical exteriority, the Other. For it is in fleshy affection that the self comes to experience the Other. The self's a-theoretical affective experience of the Other, its exposition, is at the same time the revealing of its vulnerability, which happens through sensibility (1978). Now, against classical phenomenology, Lévinas argues that sensibility is not intuition or receptivity: it is *contact,* the contact between the self and the Other. "Sensibility is not an empty box in which perceptions are housed; its original meaning is rather woven into a sensibly lived subjectivity. This body-sensibility is the possibility of non-indifference," the possibility of ethics, the sensing "of suffering pain, illness and fatigue for the other" (Farías Gutiérrez, in press). Lévinas contends that

> It is because subjectivity is sensibility—an exposure to others, a vulnerability and a responsibility in the proximity of the others, the-one-for-the-other ... —and because matter is the very locus of the for-the-other ... that a subject is of flesh and blood, a man [sic] who is hungry and eats, entrails in a skin and, thus, capable of giving the bread out of his mouth, or giving his skin. (Levinas, 2006, p. 77)

As we see, in Lévinas's ethical analysis, and against all standard phenomenological descriptions of the body, we are humans of flesh and blood *because* of our sensible subjectivity, *because* we are bearers before anything else of a disposition to alterity. Hence, it is our vulnerability and responsibility—those ethical features that predate consciousness and cognition—that make us capable of giving the bread out of our mouth or giving our skin. As Novotný remarks,

> To have a body, to be a body, to be somewhere, all this is only possible as a consequence of vulnerability, which certainly concerns physical bodies and their biological needs, but which does not come from life itself, nor from the *conatus essendi*, but from the event of exposure to the other ... [that is,] from the ethical relationship. (Novotný, 2016, p. 127)

In Lévinas, "the ethical subject is a sentient subject not a conscious ego" (Critchney, 1999, p. 188). Lévinas's insistence on the impossibility to conceptually grasp the Other, that our experience of the Other is "the excess of Being over the thought that wants to contain it" (Faría Gutierrez, in press), that the fleshy experience of the Other is not a cognitive experience, leads to an interesting view of the language of ethics—a kind of paradox.

> The paradox here is that what this ethical language seeks to thematize is by definition unthematizable, it is a conception of the subject constituted in a relation to alterity irreducible to ontology, that is to say, irreducible to thematization or conceptuality. (Critchney, 1999, p. 184)

In this ethical language we must try to say that which cannot be said. For Lévinas, this language moves within a sphere of signification that is of the order of the sensible—one that is not turned towards the world of idealities—i.e., knowledge. It is an order that is described in terms of proximity, where proximity is the signification of sensible that "does not belong to the movement of cognition" (Levinas, 2006, p. 61). This new realm of signification is what led him to distinguish between the Said and the Saying (*le Dit* and *le Dire*).

5.3 *The Said and the Saying*

The distinction between the Said and the Saying is Lévinas's effort to explore a semiotic realm where signification moves beyond the enclosures of cognition. To do so, he seeks a "language" that does not remain closed in on itself, wandering relentlessly within its own confines, naming, fixing, and reifying things, as does the language of cognition. The language of cognition is the Said. "The Said represents the movement in which there is meaning only when it is supported by the linguistic system or the ontological game that expresses it" (Cozza Sayão, 2011, p. 117). The language of the Said

> can ... be conceived as a system of nouns identifying entities, and then as a system of signs ... designating substances, events and relations by substantives or other parts of speech derived from substantives, designating identities—in sum, *designating*. (Levinas, 2006, p. 40, original emphasis)

Now, Lévinas seeks something very different: an ethical language—a language where feeling and living would come to the fore, no longer subjugated and tied to the act of understanding. He wants to unveil a more fundamental realm of signification subtly at work in our engagement with others. As Cozza Sayão puts it,

> Levinas speculates about the possibility of another context of *signification*. An alternative context of meaning to human life in which ... it is

> possible to think of signification as an *approximation* to the world and people, as an immediate opening. The idea is to think beyond the rules of Greek tradition. (Cozza Sayão, 2011, p. 109, original emphasis)

The Saying is Lévinas's response to this conundrum. The expressiveness of the Saying and how it is expressed (that is, its semiotic materiality) cannot be found *in* the linguistic word, in the diagram, in the sign, but elsewhere. The Saying operates in a different manner from the Said—although, as we shall see, it is not completely independent of the Said.

To explain the differences between the Said and the Saying Lévinas resorts to two terms: *thematisation* and *exposition*. In thematisation, we place ourselves within a conceptual system. We say things, we thematise them according to some conceptual categories. We move here within the world of cognition.

However, Lévinas notes, there is also another language that does not name, and rename, and refer things (entities) to other things in a tireless manner; there is also a language that exposes the "silent resonance" of beings and things (Levinas, 2006, p. 40). This is the language of the Saying, where we signify for the Other; we signify our proximity, approach, exposure. This exposure has a sense radically different from thematisation. "The *one* is exposed to the *other* as a skin is exposed to what wounds it" (Lévinas, 1978, p. 83, original emphasis). In this exposure, "human spiritual existence can be affirmed as sensibility and vulnerability of the flesh" (Cozza Sayão, 2011, p. 116).

The exposure signified in the Saying is a communication between the one and the other, but it should not be seen as an exchange of ideas. Communication in the Saying is a *sentient communication* between beings, a manifestation of solidarity in the presence of beings: "Saying is communication, to be sure, but as a condition for all communication, as exposure. Communication is not reducible to the phenomenon of truth and the manifestation of truth" (Levinas, 2006, p. 48). Hence,

> What comes about in Saying is an approach: I approach the Other. An approach and exposure to the Other … in a language that does not thematize, that does not make noise, that does not have time for theoretical lucubration, since the relationship that takes place there is only ethical sensitivity. (Farías Gutiérrez, in press)

In short, "To say is to approach a neighbor" (Levinas, 2006, p. 48), in a language (the language of the Saying, the language of ethics) that precedes and always goes beyond the language of the Said (the language of the conceptual).

The idea that the significations of the Saying occur before and beyond the logical-thematic significations of the Said, does not amount to negating a relationship between both. What is said in the Said receives, takes up, and echoes what is meant in the Saying, but the latter is not simply replaced with the former. Lévinas talks about *absorption.*

> The said in absorbing the saying does not become its master, although by an abuse of language it translates it before us in betraying it. The unravelling of the plot of saying does not belong to language qua said, does not belong to the last word. Saying signifies without stopping in the said. (Levinas, 2006, p. 190)

In the words of the Said reverberate the significations of the Saying. Things, their beings (essences) are not simply translations in the Said. The Said does not exhaust Being; what we can say in the language of the Said is only one of the modalities of Being. The reason is that there is a realm of Being that language and cognition are unable to fully grasp and signify.

Referring to the words that we utter in the language of the Said, Cozza Sayão notes that

> There is always something that goes beyond and transcends words, leaving out what is babbled. It is as if everything that is said about love is still insufficient and too small to bear it. That is why poetry hardly dares to touch it, that is why art never intends to represent it. (Cozza Sayão, 2011, p. 118)

In the language of the Said there is an inscription, a sign, or a sound that, by appearing, by resounding, tries to fix the "essence" or the Being of things. In the language of the Saying there is no inscription to be seen or sound to be heard; what there is, is the vibration or resonance of Being.

In the Said, the essence that resounds is on the verge of becoming a noun. "In the copula *is* scintillates or sparkles an ambiguousness between the essence and the nominalized relation" (Levinas, 2006, pp. 41–42). The language of the Said is always inadequate vis-à-vis Being. As soon as we utter a word, as soon as we make a pointing gesture, the essence that the word or the gesture (or both!) was supposed to reveal escapes, like water through one's fingers.

6 The Indispensable Task of (Mathematics) Education

As we have seen in the previous section, Lévinas makes an attempt to rethink ethics from a perspective that runs against the legendary egocentric and logocentric position of the subject-object relationship of Western thought. He turns with penetrating and accusatorial eyes to what has been the focus of this intellectual tradition since antiquity: knowledge and reason. In this sense, Lévinas joins the line of postmodern thinkers disenchanted "with the reign of reason" (Min, 1998, p. 577). He distinguishes himself from these disenchanted thinkers by the effort he makes to de-centre the subject and (in the footsteps of Buber, 1958) by the emphasis he puts on the subject-subject relationship. His work is certainly bold and daring. It emerges like a cry that comes from the deepest sub-basements of our acquired habits of thought to claim the value of human dignity.

> Levinas' protest is no doubt one of the most radical against the destruction of human dignity ever mounted by a philosopher. His protest is based on a stringent critique of the very roots of such destruction, the logo centrism or the priority of reason, logic, idea, concept, and the categories of immanence that pervade the Western philosophical tradition. (Min, 1998, p. 572)

He endeavours to extract from oblivion a dimension that goes beyond Being as conceptualised in Western ontology and to show the existence of a realm where Being signifies otherwise: what Being signifies in a language that does not make noise is our central relationship with the Other. In the new realm of signification, language does not operate in the technical, epistemological manner that we find in the work of Plato and Aristotle, Étienne Bonnot de Condillac in the 18th century, or in Edmond Husserl in the 20th century. Lévinas reveals a substratum of signification that is at work, tacitly, in our encounter and proximity of the Other. In this proximity our conceptual epistemological categories and mechanisms are put on hold and we encounter the Other as *is*. "The Other is appreciated precisely *as* Other, in her radical alterity and irreducible singularity, only when thought renounces its totalitarian hubris and learns to think of the Other on her own terms … 'beyond essence'" (Min, 1998, pp. 573–574, original emphasis).

Yet, the Other that we encounter in its angelical presence has been bereft of its constitutive social, political, and cultural nature. It is an Other that, through its presence, covers us with the light of its epiphany. We encounter the Other in an *unmediated* way, directly, a sublime encounter of two souls. The whole of the constitutive cultural-historical mediating relations vanishes into thin air and we end up encountering the Other as a historical abstraction. "The Other of Levinas' is 'concrete' only in the phenomenological sense of the lived immediacy of the face-to-face encounter" (Min, 1998, p. 580).

> Levinas finds a corrective of this egocentrism in heterocentric subjectivity, the ego questioned and challenged by the irreducibly transcendent claim of the Other. The ethical center has shifted from the ego to the Other. Levinas' ethical metaphysics, however, remains an essentially Cartesian philosophy of the subject, albeit with a different ethical content. It still remains a philosophy of the subject as individual, without intrinsic socio-historical mediation … The individualist subjectivism of modern philosophy is only abstractly negated … Levinas' metaphysics remains infected by the very subjectivism it has not wholly overcome. (Min, 1998, p. 584)

A genuine encounter with the Other, I would argue, cannot be achieved without taking into account our earthly conditions. Our encounter with the Other must occur at the very heart of those conditions, in our worldly and quotidian contexts, in concrete, material life where, unfortunately, people oppress, marginalise, discriminate against, and control the Other. Ethics should not be a mere encounter of self and Other but a *liberating* force that can help us come to grips with the historical, political, and economic sources and structures of oppression, violence, and inequality. This is why I prefer to think of ethics *as liberation*. In such an ethics we need to bring to the fore a pluralistic sense of collectivity and discuss together what a common good could be. An ethics of liberation "implies *denouncing* how we are living and *announcing* how we could live" (Freire, 2004, p. 105, original emphasis). It should recognise the political, economic, and societal structures that frame our life. It should help us imagine and implement the collective actions required to transform these structures.

It is here where education has a place in society. Indeed, education cannot continue to be reduced to the reproduction of society in the benefice of some groups or classes and the reproduction of inequality (Bourdieu & Passeron, 1990). Now, if a profound societal change has to occur, education has to be part of the equation. Education is a vital pivot in the emergence of new forms of alterity and new conceptions of society. Education should be part of what Marx and Engels called the "rearrangement of the empirical world" for individuals to experience the really human dimension of our existence.

> If man [sic] draws all his knowledge, sensation, etc., from the world of the senses and the experience gained in it, the empirical world must be arranged so that in it man experiences and gets used to what is really human and that he becomes aware of himself as man … If man is shaped by his surroundings, his surroundings must be made human. If man is social by nature, he will develop his true nature only in society, and the power of his nature must be measured not by the power of separate individuals but by the power of society. (Marx & Engels, 1956, p. 176)[5]

Sociologist Michel Freitag complained that nowadays there is no longer a social movement that produces an intelligibility of collective existence:

> no [social movement] develops a new conception of the whole order [...]. No [movement] produces a new general intelligibility of society and history, collective existence, its purposes and its constraints. In a word, there is no longer any politics. Politics, as a reflexivity of living together (this time around the planet, and this time as knowing how to live with techniques), and politics as collective responsibility for the ends of social life, needs to be reinvented. (Freitag, 2002, pp. 243–244)

It is precisely in the school that the foundations of these political social movements should begin. Education should be the site to imagine new possibilities and new more dignifying futures (D'Ambrosio, 2004, 2006). And ethics must be part of this emancipatory educational project.

In this view, the societal justification of the school no longer rests on the preparation of the child to enter the workplace. If it is true that the school cannot be disentangled from the economic sphere of society, it is even truer that the school cannot and should not be limited to the economic dimension of society, as is the case today (Radford, 2018d, 2020e). So, in the view that I am defending here, the reason parents send their children to school each morning is no longer because in the school their children will find the knowledge that they require to find a job. The justification—the *raison d'être*—for children to cross the threshold of the school each morning should be because it is in school that children can experience something that they *cannot* in society: it is in school that children could make the experience of democracy, of collective life, solidarity, and inclusivity.

However, the conditions for the emergence of new forms of alterity and new conceptions of society and the school cannot appear out of the blue. We need to explore fresh paths towards the renewal and imagination of new forms of relating to others and working with others. In the theory of objectification, we attempt to do that through the practice of what we term a *communitarian ethics.*

7 Towards a Communitarian Ethics

As stated in Chapter 2, the theory of objectification seeks to offer a perspective where learning appears as a collective endeavour: knowledge is collectively produced by teachers and students in what we have termed *joint labour*. In joint labour the mathematics classroom is not witness to two parallel processes (teaching on one hand and learning on the other) but a single process: the process of teaching-learning. Joint labour does not mean that teachers and students do the *same* thing. There is a division of labour. But this division of labour does not preclude teachers and students from working *together*, hand in hand.

Now, not all interactional classroom activity can be seen as an instance of joint labour. It happens almost systematically that when, in seminars, I talk about joint labour and show the phases of joint labour (see Figure 17 in Chapter 5), I am told that this is what people have been doing for over two decades now (see Radford, 2020f). The misunderstanding is that classroom interaction is seen *ipso facto* as joint labour. In joint labour, teachers and students interact among themselves, but they do so in a *strong* social sense. Interaction in joint labour does not simply mean a series of actions carried out by a student; nor does it mean a coordination of actions by a group of students (regardless of how complex such a coordination may be). Engaging in *joint labour* is much more than merely interacting—much more than solving problems in groups. The etymological-conceptual roots of joint labour, as I mentioned in previous chapters,[6] are to be found in the German term for activity, *Tätigkeit*, which carries with it a specific *ontological* meaning—the fundamental idea that "Human nature constitutes itself socially and historically" (Freire, 2004, p. 105). Through the *inter*-actions that occur in *Tätigkeit* the individuals produce things to fulfill their needs. However, this "production must not be considered simply as being the reproduction of the physical existence of the individuals" (Marx, 1998, p. 37). It does not merely mean getting together to get something done. The *inter*-actions of the individuals in joint labour (*Tätigkeit*) include, in a decisive manner, modes of social, cultural, and historical production. Through these means, individuals produce *collectively* and their production is *not alienating*.

Alienation does not consist in that the individuals' labour becomes an object, nor is it that the individuals' lives become crystalised in the object (Fischbach, 2012). It is, indeed, in the nature of labour to produce something and that the individuals express and recognise themselves in it. Alienation consists instead in the precise fact that the produced object is no longer the individuals' expression. What is alienating here is hence *the loss of expressivity of life in the object*, the loss of self-expression in the object. Instead of expression, achievement, and self-realisation, we have a product that becomes a *thing*. In direct teaching, the students do not recognise themselves in the mathematical ideas that are produced in the classroom. The students are powerless. They are alienated from the process of production and from the resulting product. But, as I argued in (Radford, 2014b), the problem is not overcome by giving the power to the students. This strategy creates another kind of alienation (for a detailed discussion of alienation in mathematics classrooms see Radford (2012b, 2013b, 2016b) and Radford and Roth (2017)).

What would non-alienating production look like? Non-alienating production would not be producing when we do not find realisation in what we do, nor would it be producing when we have lost control of our lives. In a marginal note, Marx writes:

> Let us suppose that we had produced as human beings. In that event ... in the individual expression of my own life I would have brought about the immediate expression of your life, and ... I would have directly *confirmed* and *realized* my authentic nature, my *human, communal* nature. Our productions would be as many mirrors from which our natures would shine forth. (as cited in Sayers, 2011, p. 99, original emphasis; see also Marx, 1968, p. 33)

Non-alienating production is based, as we see, on a labour process that is not only an expenditure of energy, but at the same time a profound *relationship with the other*: self and other *affirm* him/herself doubly in their production.

A similar idea is elaborated by Leont'ev in the following passage dealing with labour (non-alienating collective labour):

> Labor is a process ... that is realized ... under conditions of people's joint activity, under conditions of a human collective, and ... in a social, that is, collectively expressed way. Through this process, people enter into communication with one another. It is not so much a matter of communication that is primarily verbal, of course, but of communication in the sense of participation in a joint action, in the sense of participation in the process of labor, first and foremost. (Leont'ev, 2005, p. 60)

In joint activity or joint labour, communication, as Leont'ev notes, is not just talking nor is it just naming things. There is behind the communicative act something truly social—a *committing stance* that participants adopt to work together to achieve something communal.

To sum up, in joint labour (*Tätigkeit*), people produce in a collective, non-alienating manner. That is, they express themselves in what they produce, and find at the same time human realisation in what they produce and in the process of its production.

Now, neither Marx nor Leont'ev says so, but the underpinning committing stance required in non-alienating labour/learning is inevitably of an *ethical* character.

In the theory of objectification, the ethics that we are interested in is neither of a Hobbesian nor of a Kantian nature. The ethics in which we are interested is a context-sensitive, fluid, personal, and cultural sensible *relation* of *responsibility* between self and others (Giroux, 2005). It is an ethics turned towards the reflexive and critical constitution of what Marx (1988, p. 108) called "human powers," such as will, love, cooperation, and solidarity—i.e., capacities that affirm the social, cultural, and historical nature of individuals and where our relations to others as sentient beings becomes the ontological condition of our existence. This ethics lies on the recognition that our historical, cultural, and material origin embeds and refracts dynamic and antagonistic visions and conceptions of the world and of what a good life can mean. It is the vitality of contradictions (Bruce, 2013) that gives substance to social human life, always changing, always challenged, and that makes meaning a polyphonic event, always arising and evolving "in the context of struggle" (Juzwik, 2004, p. 540).

Let's go back to the example with which I started this book: The Piggy Bank example. What we see in that example (but this is true of all other examples discussed in other chapters as well) is that the teacher is not teaching in the transmissive sense. Neither is the teacher merely observing the students produce their own knowledge. The teacher works *with* and *for* the students, *and reciprocally*. In Figure 3 and the associated dialogue (Chapter 1), we see, indeed, that instead of saying how to see the terms of the sequence, the teacher *invites* the students to find a new way of perceiving the terms: a way that rests on identification of the mathematical variables and their relationship.

29. Mrs. Giroux: (*Trying to make noticeable to the students the co-variational structure*) What do you remark about Week 5 (*She shows the glass corresponding to Week 5; see Figure 48, Picture 1*) and (*Pointing to the red bingo chips*) the number of bingo chips? (*Making the same actions*) The fourth week and the number of bingo chips? (*See Figure 48, Picture 2*)

In engaging with the students in this manner, the teacher assumes *responsibility*. In doing so, she *exposes* herself. She *appears* in the social scene, occupying a space with her body and her actions. She positions herself and becomes a *presence*. And by becoming a presence, she makes herself *vulnerable*. She *assumes the risk* that things may turn bad. For instance, the students can get frustrated. But she decides to go ahead. It would have been much easier to simply tell the students what to do! But then, the teacher and the students would have walked the path towards alienation. The teacher would have imposed on the students a regime of truth—a way of doing mathematics. Her *invitation* makes room for the students to engage in a cultural-historical way of thinking about sequences. Now, this invitation is more of an enticement to look at this way of mathematical thinking, to encounter it, to examine it, to discuss it. The teacher's invitation moves *at the same time* on another level. *Her invitation is more than a simple cognitive call.* It is an ethical call to the Other.

Indeed, in grasping the glasses of Week 5 and Week 4 the teacher tries to *reach* the students and *engage* them in the discussion. But there is something more to the teacher's ethical call here that is important to note: The unfolding ethical relationship that appears through the teacher's call is a relationship that depends crucially on the *body*. It is an *embodied ethical relationship* that is manifested in the touching of the glasses, in the teacher's posture, in the tonality of the uttered words, in the gestures that point to the material objects, and in the rhythm that provides coherence to the various facets of embodied expression.

The students may decide to ignore the teacher's call and to simply ask the teacher directly how to solve the problem, without further ado, or hide behind the wall of "I do not understand." However, the students decided to respond to the teacher's call. They decided to engage by *responding* to the teacher's invitation. Like the teacher, the students were not sure that things would work out well. Yet, they decided to join the teacher, and to take risks too. They decided to *trust* the teacher:

30. Albert: It's always twice ...
31. Mrs. Giroux: (*Repeating*) It's always twice.
32. Krysta: It's the double of what you ... (*See Figure 48, Picture 3*) No! (*She watches the artefacts intensely for a while*) I am confused!
33. Albert: Yeah! It's twice, look! (*Counting the red chips*) 1 + 1, 2; 2 + 2, 4; 3 + 3, 6; 4 + 4 (*See Figure 48, Picture 4*).

When Krysta, in line 32, says "It's the double of what you ...," she offers the others a new possibility to see the mathematical meaning of the bingo chips. Yet, the idea is not completely clear; it lacks coherence. Krysta acknowledges it by saying after a pause "I am confused!" But she allowed Albert to perceive something that so far had escaped his attention. In line 33 he says "Yeah!" There is a multiplicative relationship in the variables that start appearing on their conceptual horizon. The responsibility, exposition, vulnerability, and trust that are embedded in this passage are part and parcel of the ethical dimension of the joint labour—an extremely emotional dimension.

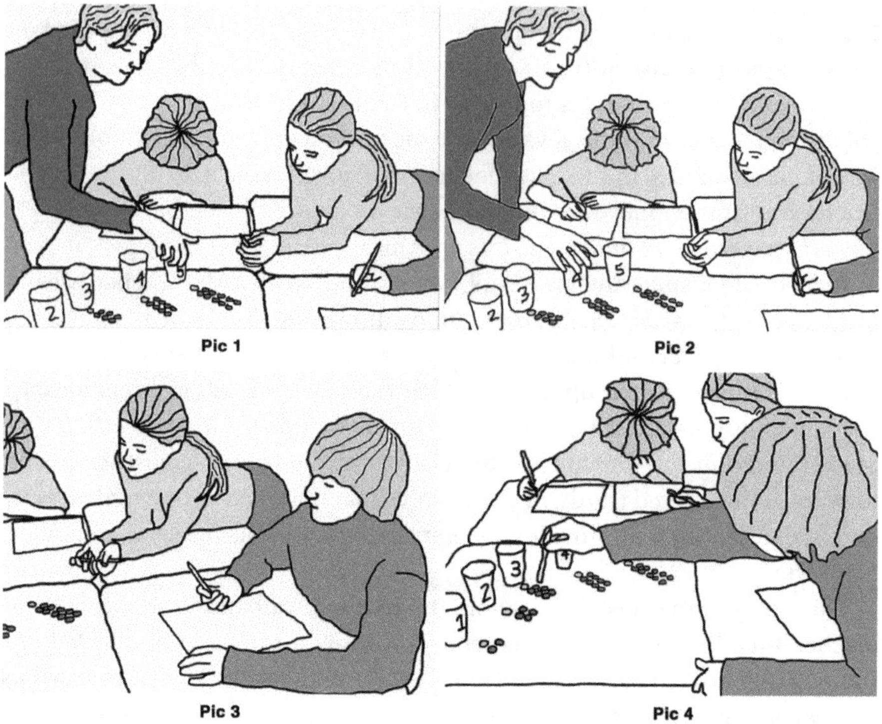

FIGURE 48 The unfolding ethical relationship between the teacher and the students depends crucially on the body

From the joint labour of the teacher and the students, algebra is being materialised in a process that is dynamic, sensible, rhythmic, concrete, linguistic, symbolic, and corporal. What the teacher and students are doing is producing a *common work* (*oeuvre commune*): the materialisation of mathematical knowledge that, as it comes into being, invades the classroom, like music would invade a concert hall. Ethics appears here in the manner in which the teacher and the students *commit* to their joint endeavour, in *how* they engage, respond to, take care of, and relate with, each other. This ethics orients the collaboration and mathematical inquiry between teachers and students. It makes the teaching-learning activity a sensible and intellectual, critical and poetic experience. Through this ethics a deep understanding of mathematics is made possible while at the same time making this understanding a collective, social, and culturally enriching adventure for both students and teachers.

This ethics, that in the theory of objectification we term *communitarian ethics*, is centred on three elements: a) responsibility, b) commitment, and c) the care for others. These three elements orient joint labour and come to configure the essential structure of subjectivity.

a) *Responsibility* appears here as an unconditional undertaking towards others. It realises a union, a link, a connection, and a bond between self and others. Responsibility is expressed in the answerability we make to the call of the Other, a call that comes not necessarily from a linguistic or semiotic formulation, as Lévinas shows, but from the mere presence of what we ourselves are not. Responsibility is the answer to what the Norwegian philosopher Kund Løgstrup (1997) calls the *ethical demand.*

b) *Commitment* is both a promise and its realisation of doing everything possible to work side by side with others in the course of our joint labour (e.g., trying to understand, trying to contribute to the common work).

c) *The care for others,* far from being an act of condescendence, or a patriarchal act, or simply caring for someone, is a "relational involvement" (Edwards, 2009, p. 234); as such, it needs no justification. It is a way of being with another and being-for-other. The care of the Other requires the constitution of those sensibilities or human capacities to which Marx (1988) refers, in particular the sensibility of the attention and recognition of others and their material and spiritual needs. Although caring for the Other opens up the possibility of seeing ourselves in the Other; of recognising our vulnerability in the vulnerability of the Other, the importance of caring for the Other is to go beyond ourselves, to ourselves be dragged powerfully into the world and to position ourselves there, with-the-Other.

But, of course, these elements of the communitarian ethics are not the product of natural evolution. They are elements of the form of otherness that have been historically produced. If the elements of the communitarian ethics were the business of natural evolution, we would not be living in the world we are in.

In recent work with 9–10-year-old students of primary education in a public school in Bogota, Colombia, we observe that in their classroom forms of human collaboration and knowledge production, the students resort to what may be termed "closed ethics"—one where the relationship to otherness is alienating: it is an ethic of superior/inferior, potent/impotent, knower/ignorant, authority/vassal. The students conceptualise the teacher as the possessor of knowledge and power. Furthermore, the students conceptualise themselves as submissive to the teacher and her knowledge, even when the teacher tries to conceptualise herself differently and, to encourage student interaction, organises the classroom into small groups. The students configure enclosed small groups and erect aggressive barriers between them and the others—a clique culture (Radford & Lasprilla Herrera, 2020).

As bell hooks notes, often "students are unwilling to acknowledge that their minds have been colonised, that some have been learning to be oppressors

[and] to dominate," while others have been learning to "passively accept the domination of others" (hooks, 2015, pp. 177–178). As the doctoral research conducted by Lasprilla Herrera suggests, students find the move to an ethics of solidarity terribly threatening (Lasprilla Herrera, Radford, & Léon, in press).

I do not believe that ethics can be formally learned. We can talk about it, and so on. But an emancipatory ethics of liberation cannot appear as an imposition, which would defeat its purpose. It can only appear through its daily practice, in our acts and deeds "to support the real transformation of the conditions of life of all the individuals" (Rubinstein, 1983, p. 402). Ethics can appear as an invitation to put into question our habits of thought and conduct—what in Chapter 9 was called, following Scott, the Question of ethics, "an interruption in an ethos, an interruption in which the definitive values that govern thought and everyday action lose their power and authority" (1990, p. 4) an interruption that requires us "to become alert to exclusions and to forgotten aspects in a people's history, to overhear what is usually drowned out by the predominant values, to rethink what is ordinarily taken for granted" (Scott, 1990, p. 7).

An emancipatory liberating ethics would require us to work towards the creation of the conditions of possibility for such an ethics to emerge. This task implies changes not only in the students but in the educators as well:

> It is no longer the educator who deposits all his or her knowledge in the student, and the latter who passively receives from the former a knowledge that is given to him or her, but, in a critical position, the student takes the floor and the educator allows the dialogical relationship as a place for the liberation of both [teacher and student], given that in the circulation of the word itself the subject becomes desalinated, making the encounter with the other a possibility of social creation and not a persecutory moment of the struggle of all against all. (Fernández Mouján, 2016, p. 103)

8 Synthesis

In the first section of this chapter I suggested that all teaching-learning activities have a basic composition: they are based on two interwoven axes: (1) the axis of the *forms of human collaboration*, and (2) the axis of the *forms of knowledge production*.

Then, I argued that both axes include an ethical component. In the first one, ethics appears related to the manners in which classroom participants (the teacher included) relate to others. In the second one, ethics appears related to the way knowledge is legitimised. Ethics appears as soon as some procedures of inquiry, proving, problem solving, etc., are highlighted, preferred, etc., to the detriment of other procedures. I showed how direct teaching and constructivism resort to two different ethics. Then, I made an incursion into three of the most important ethical systems of the Western intellectual tradition: Kant's, Hobbes's, and Lévinas's. The interest was to show that behind these ethical systems lies a concept of the individual and her relationship to society. In my short analysis it became apparent that Kant's and Hobbes's systems work around a view of the social as a conglomerate or an assemblage of individuals. Kant's ethics stresses a commitment to reason considered as a human universal faculty (a legislative reason as understood by the European cultural movement of the 18th century: The Enlightenment). It also stresses the value that was given to autonomy and freedom as the chief features of what it means to be human. Drawing on these ideas Kant formulated what he called the categorical imperative, where the Other is simply assimilated to the self. Hobbes, by contrast, in a more pragmatic mood, articulated an ethics as contract—a contract that tries to balance the multitude of interests of the individual and that reflects perfectly the individualism of his time. Lévinas erupts onto the ethical scene with a different project, one that is erected against the ashes of some of the major cultural and political tragic disasters of the first half of the 20th century: the World Wars and the Holocaust. His ethical project is built against two of the most cherished values of Western thought: knowledge and reason. But not knowledge and reason *per se*; rather knowledge and reason as weapons in the domestication, manipulation, annihilation, and obliteration of the Other. Lévinas offers us a healing ethics from where we could recover what we have lost: human dignity. His ethics is an invitation to think differently, to renew our approach to the Other, to embrace the Other as it is, unconditionally, by what it is: a *presence* (or a *face*, as Lévinas says again and again) to which *we* owe our own existence. His ethics is rich, sublime, and magnificent. It offers us the possibility to uncover a new realm of signification, a realm without words, before and beyond logos or discourse, a language without signs other than those we need to live for-the-other. But, following in the steps of the very same intellectual tradition he tried to overcome, the Other in his ethical system remains an abstract entity, deprived of all the cultural-historical-political mediations that are at work in life every minute, every second, silently, affecting—most of the time, without our noticing—the ways we relate to others. Driven by the current capitalist forms of production, our contemporary life is such, indeed, that human activity has become based on material advancement to the detriment of a genuine social dimension. Deploring the sad state of contemporary societies, Mészáros notes that to comply with the capitalist requirements,

> the legitimately feasible objectives of human activity *must* be conceptualized in terms of material advancement through the agency of the natural sciences, remaining blind to the *social dimension* of human existence in other than essentially functional/operative and manipulative terms. For an alternative view would necessitate abandoning the "standpoint of political economy," equivalent to the vantage point of capital, which must see even in living labour nothing but a "material factor of production." (Mészáros, 2010, p. 29, original emphasis)

The lack of attention to the whole of the constitutive cultural-historical mediating relations that permeate our lives in Lévinas's ethics makes the encounter with the Other a historical and political abstraction.

I pleaded then for an ethical project that recognises the political, economic, and societal structures that frame our life and that would help us imagine and implement the collective actions required to transform these structures. I argued that such an ethical project should be at the centre of education in general, and mathematics education in particular.

In the rest of the chapter, I sketched the ethics of the theory of objectification—the *communitarian ethics.* Communitarian ethics is an attempt to explore fresh paths towards the renewal and imagination of new forms of relating to others and working with others. It is in tune with the general tenets of the theory of objectification and its understanding of learning as a collective endeavour. It is also in tune with the educational project in which the theory is embedded, an educational project that I mentioned in Chapter 1 and that is worth recalling in these closing remarks:

> The theory of objectification posits the goal of Mathematics Education as a political, societal, historical, and cultural endeavour aimed at the dialectical creation of reflexive and ethical subjects who critically position themselves in historically and culturally constituted mathematical discourses and practices, and who ponder new possibilities of action and thinking.

Communitarian ethics plays a fundamental role in the central concept of the theory, namely *joint labour*, the joint labour of teachers and students to the extent that, without taking the communitarian ethics into consideration, joint labour does not make sense.

Communitarian ethics draws on a dialectical-materialist concept of the self that appears throughout the book and that I articulated explicitly in Chapter 9. It is a concept of the self that acknowledges the cultural-historical nature of the human, the irradicable political nature of education, and its crucial role in the transformation of society (Freire, 1970, 2016; Vygotsky, 1929, 1997b). Communitarian ethics draws also on Lévinas's ethics, yet not Lévinas's ethics as such. It draws on Lévinas's ethics once we have done to it what Marx did with Hegel's dialectics; that is, once we have turned it upside down so that it ends up removed from its pervasive idealism, standing now in the terrestrial world of concrete beings enmeshed in the political, economic, and cultural processes of society. In this light, joint labour, and with it teaching and learning, or rather teaching-learning (*obuchenie*, to use Vygotsky's word) appears much more than social interaction. It is the bearer of multiple and contested threads of history and culture, of tension and conflict. *Joint labour rests on the constant practice of communitarian ethics*. As such, joint labour is the bearer of hope. Not of hope in a transcendental, metaphysical, or romantic sense, but hope in the most terrestrial and political sense:

> Because we find ourselves subjected to countless limitations—difficult obstacles to overcome like the dominant influence of fatalist understandings of history or the power of neoliberal ideology, whose perverse ethic is founded in the laws of the market—there has never been a greater need to underscore educational practice with a sense of hope than there is today. (Freire, 2004, p. 100)

Hope in this Freirean sense means the hope that *together*, we can make the world that we inherited and for which we are responsible, a better place: a dignifying world for those to come and live as humans: that is, in peace and in solidarity. Mathematics education has a tremendous role to play here. If not, what would be its point?

Acknowledgement

Part of this chapter draws on Radford, L. (2021). La ética en la teoría de la objetivación. In L. Radford & M. Silva Acuña (Eds.), *Ética: Entre educación y filosofía* (pp. 107–141). Bogota: Universidad de los Andes. Used with permission of the copyright holder.

Notes

1 I use the term "form" (*Gestalt*) in its dialectical materialism sense: just as there are *forms* of property and *forms* of society, there are *forms* of human interaction and *forms* of inquiry. They are the very expressions of the social, cultural, and historical background that frame them; they are the ones that are materialised in the relationships between subjects and are evident in the immediacy and banality of everyday life.

2 The elements of these axes are related to what in Chapter 9 I termed *Semiotic Systems of Cultural Signification*.

3 In a visit to Chile a few years ago, I was told the story of a constructivist teacher who wanted so desperately to help a student by offering hints on how to solve a mathematical problem. But despite his urgent desire to do so, he proudly restrained himself. He acted ethically.

4 In general, I will use Lévinas (with "é"). However, I will also use Levinas, depending on the spelling in the cited source.

5 The meaning of what is really human, of course, is not something determined beforehand. It is something always to be disputed, to be challenged.

6 See also Radford (2020b, 2020f).

References

Adorno, T. (1998). *Beethoven. The philosophy of music*. Stanford, CA: Stanford University Press.

Adorno, T. (2006). *History and freedom*. Cambridge, England: Polity Press.

Alcoff, L. (1988). Cultural feminism versus post-structuralism: The identity crisis in feminist theory. *Signs, 13*(3), 405–436.

Althusser, L. (2005). *For Marx*. London, England: Verso.

Amariglio, J., Resnick, S., & Wolff, R. (1988). Class, power, and culture. In C. Nelson & L. Grossberg (Eds.), *Marxism and the interpretation of culture* (pp. 487–501). London, England: MacMillan Education.

Anderson, C. W. (1993). The modulation of feeding behavior in response to prey type in the frog rana pipiens. *Journal of Experimental Biology, 17*, 1–12.

Arendt, H. (1958). The modern concept of history. *The Review of Politics, 20*(4), 570–590.

Arendt, H. (1992). *Lectures on Kant's political philosophy*. Chicago, IL: The University of Chicago Press.

Arievitch, I. (2003). A potential for an integrated view of development and learning: Galperin's contribution to sociocultural psychology. *Mind, Culture and Activity, 10*(4), 278–288.

Aristotle. (1984). *The complete works. Metaphysics* (Vol. 2, W. D. Ross, Trans.). Princeton, NJ: Princeton University Press.

Aristotle. (1992). *The politics* (T. A. Sinclair, Trans.). London, England: Penguin Books.

Artigue, M. (1995). The role of epistemology in the analysis of teaching/learning relationships in mathematics education. In Y. M. Pothier (Ed.), *Proceedings of the 1995 annual meeting of the Canadian mathematics education study group* (pp. 7–21). London, Canada: University of Western Ontario.

Artigue, M. (2009). Didactical design in mathematics education. In C. Winslow (Ed.), *Nordic research in mathematics education. Proceedings of NORMA08*. Rotterdam, the Netherlands: Sense Publishers.

Artinian, A. (2017). Radical currents in Soviet philosophy: Lev Vygotsky and Evald Ilyenkov. *Socialism and Democracy, 31*(2), 95–121.

Arzarello, F. (2006). Semiosis as a multimodal process. *Revista Latinoamericana de Investigación en Matemática Educativa* (Special Issue on Semiotics, Culture, and Mathematical Thinking, Guest Editors: L. Radford & B. D'Amore), 267–299.

Arzarello, F., Bazzini, L., & Chiappini, C. (1994). *L'algebra come strumento di pensiero, Analisi teorica e considerazioni didattiche* [*Algebra as a tool for thought, theoretical analysis and didactic considerations*]. Pavia, Italy: Dipartimento di Matemática, Università di Pavia, Progetto Strategico del C.N.R.

Asper, M. (2009). The two cultures of mathematics in ancient Greece. In E. Robson & J. Stedall (Eds.), *The Oxford handbook of the history of mathematics* (pp. 107–132). Oxford, England: Oxford University Press.

Atkins, K. (2005). *Self and subjectivity*. Oxford, England: Blackwell.

Atkinson, J. (2000). *The developing visual brain*. Oxford, England: Oxford University Press.

Atweh, B. (2014). Mathematics education and democratic participation between the critical and the ethical: A socially response-able approach. In O. Skovsmose & B. Greer (Eds.), *Opening the cage: Critique and politics of mathematics education* (pp. 325–342). Rotterdam, the Netherlands: Sense Publishers.

Atweh, B., & Brady, K. (2009). Socially response-able mathematics education: Implications of an ethical approach. *Eurasia Journal of Mathematics Science and Technology Education, 5*(3), 267–276.

Atweh, B., & Cooper, T. (1991). Hegemony in the mathematics curricula: The effect of gender and social class on the organization of mathematics teaching for year 9 students. In F. Furinghetti (Ed.), *Proceedings of the 15th conference of the international group for the Psychology of Mathematics Education* (pp. 88–95). Assisi, Italy: PME.

Atweh, B., Forgasz, H., & Nebres, B. (2001). *Sociocultural research on mathematics education*. Mahwah, NJ: Lawrence Erlbaum.

Bacon, F. (1906). *Francis Bacon's essays*. London, England: Dent.

Bakhtin, M. M. (1981). *The dialogical imagination*. Austin, TX: University of Texas Press.

Bakhtin, M. M. (1986). *Speech genres and other late essays*. Austin, TX: University of Texas Press.

Bakhurst, D. (1991). *Consciousness and revolution in Soviet philosophy. From the Bolsheviks to Evald Ilyenkov*. Cambridge, England: Cambridge University Press.

Baldino, R., & Cabral, T. (1998). Lacan and the school's credit system. In A. Olivier & K. Newstead (Eds.), *Proceedings of 22nd conference of the international group for the Psychology of Mathematics Education* (Vol. 22, pp. 56–63). Stellenbosch, South Africa: PME.

Bartolini Bussi, M. G. (1991). Social interaction and mathematical knowledge. In F. Furinghetti (Ed.), *Proceedings of the fifteenth annual conference of the international group for the Psychology of Mathematics Education* (Vol. 1, pp. 1–16). Assisi, Italy: PME.

Bartolini Bussi, M. G., & Mariotti, M. A. (1999). Semiotic mediation: From history to the mathematics classroom. *For the Learning of Mathematics, 19*(2), 27–35.

Bartolini Bussi, M. G., & Mariotti, M. A. (2008). Semiotic mediation in the mathematics classroom: Artefacts and signs after a Vygotskian perspective. In L. English (Ed.), *Handbook of international research in mathematics education* (2nd ed., pp. 746–783). New York, NY: Routledge, Taylor and Francis.

Baudrillard, J. (1968). *Le système des objets* [*The system of objects*]. Paris, France: Gallimard.

Bauman, Z. (1993). *Postmodern ethics*. Oxford, England: Blackwell.

Beaud, M. (2004). *A history of capitalism*. Delhi, India: Aakar Books.

Bednarz, N., Kieran, C., & Lee, L. (1996). *Approaches to algebra, perspectives for research and teaching*. Dordrecht, the Netherlands: Kluwer.

Bernstein, B. (2003). *Class, codes and control. The structuring of pedagogical discourse* (Vol. IV). London, England: Routledge.

Bikner-Ahsbahs, A., & Prediger, S. (2014). *Networking of theories as a research practice in mathematics education*. Cham, Switzerland: Springer.

Biro, D., Sousa, C., & Matsuzawa, T. (2006). Ontogeny and cultural propagation of tool use by wild chimpanzees at Bossou, Guinea: Case studies in nut cracking and leaf folding. In T. Matsuzawa, M. Tomonaga, & M. Tanaka (Eds.), *Cognitive development in chimpanzees* (pp. 476–508). Tokyo, Japan: Springer-Verlag.

Bishop, A. (1985). The social psychology of mathematics education. In L. Streefland (Ed.), *Proceedings of the 9th conference of the international group for the Psychology of Mathematics Education* (Vol. 2, pp. 1–13). The Netherlands: PME.

Bishop, A. (1988). *Mathematics education and culture*. Dordrecht, the Netherlands: Kluwer.

Bishop, A., & Pompeu, G. (1991). Influences of an ethnomathematical approach on teacher attitudes to mathematics education. In F. Furinghetti (Ed.), *Proceedings of the 15th conference of the international group for the Psychology of Mathematics Education* (Vol. 1, pp. 136–143). Assisi, Italy: PME.

Blunden, A. (2010). *An interdisciplinary theory of activity*. Chicago, IL: Haymarket Books.

Boas, F. (1944). *The mind of primitive man*. New York, NY: Macmillan.

Boero, P., Pedemonte, B., & Robotti, E. (1997). Approaching theoretical knowledge through voices and echoes: A Vygotskian perspective. In *Proceedings of the 21st international conference for the Psychology of Mathematics Education* (Vol. 2, pp. 81–88). Lahti, Finland: PME.

Boesch, C., Marchesi, P., Marchesi, N., Fruth, B., & Joulian, F. (1994). Is nut cracking in wild chimpanzees a cultural behaviour? *Journal of Human Evolution, 26*, 325–338.

Borba, M., & Villareal, M. (2006). *Humans-with-media and the reorganization of mathematical thinking*. New York, NY: Springer.

Bostock, D. (1980). Aristotle's account of time. *Phronesis, 25*(2), 148–169.

Bourdieu, P., & Passeron, J. C. (1990). *Reproduction in education, society and culture*. London, England: Sage.

Bowers, N., & Lepi, P. (1975). Kaugel valley systems of reckoning. *Journal of the Polynesian Society, 84*(3), 309–324.

Boylan, M. (2016). Ethical dimensions of mathematics education. *Educational Studies in Mathematics, 92*(3), 395–409.

Bresson, A. (2016). *The making of the ancient Greek economy: Institutions, markets, and growth in the city-states*. Princeton, NJ: Princeton University Press.

Brousseau, G. (1978). *La Monographie d'un enfant en difficulté: l'enfant Gaël.* Bordeaux, France: IREM de Bordeaux. Retrieved from http://guy-brousseau.com/wp-content/uploads/2010/12/Monographie-d'un-enfant-en-difficulté-l'enfant-Gaël.pdf

Brousseau, G. (1988). Le contrat didactique: le milieu. *Recherches en Didactique des Mathématiques, 9*(3), 309–336.

Brousseau, G. (2002). *Theory of didactical situations in mathematics*. Dordrecht, the Netherlands: Kluwer.

Brousseau, G. (2003). *Glossaire de quelques concepts de la théorie des situations didactiques en mathématiques* [*Glossary of some concepts of the theory of didactic situations in mathematics*]. Retrieved January 20, 2007, from http://Dipmat.Math.Unipa.It/~grim/gloss_fr_brousseau.pdf

Brown, T. (2010). Truth and the renewal of knowledge: The case of mathematics education. *Educational Studies in Mathematics, 75*(3), 329–343.

Bruce, G. (2013). *The vitality of contradiction. Hegel, politics, and the dialectic of liberal-capitalism.* Montreal, Québec: McGill-Queen's University Press.

Bruner, J. (1990). *Acts of meaning*. Cambridge, MA: Harvard University Press.

Bühler, K. (2011). *Theory of language. The representational function of language.* Amsterdam, the Netherlands: John Benjamins.

Burkert, W. (1985). *Greek religion. Archaic and classical.* Maiden, MA: Blackwell.

Burkert, W. (1995). *The orientalizing revolution. Near Eastern influence on Greek culture in the early archaic age*. Cambridge, MA: Harvard University Press.

Butler, J. (1999). *Gender trouble. Feminism and the subversion of identity.* London, England: Routledge.

Cai, J., & Knuth, E. (2011). *Early algebraization*. New York, NY: Springer.

Campbell, S. (2002). Constructivism and the limits of reason: Revisiting the Kantian problematic. *Studies in Philosophy and Education, 21*, 421–445.

Cantoral, R. (2013). *Teoría socioepistemológica de la matemática educativa* [*Socio-epistemological theory of educational mathematics*]. Barcelona, Spain: Editorial Gedisa.

Cassirer, E. (1980). *The philosophy of symbolic forms: Language* (Vol. 1). New Haven, CT: Yale University Press.

Chaiklin, S. (2019). The meaning and origin of the activity concept in Soviet psychology—with primary focus on A. N. Leontiev's approach. *Theory & Psychology, 29*(1), 3–26.

Clot, Y. (2015). Vygotski avec Spinoza, au-delà de Freud. *Revue Philosophique de la France et de l'Étranger, 140*(2), 205–224.

Cobb, P. (1988). The tension between theories of learning and instruction in mathematics education. *Educational Psychologist, 23*(2), 87–103.

Cobb, P., Stephen, M., McClain, K., & Gravemeijer, K. (2001). Participating in classroom mathematical practices. *The Journal of the Learning Sciences, 10*(1–2), 113–163.

Cobb, P., & Yackel, E. (1996). Constructivist, emergent, and sociocultural perspectives in the context of developmental research. *Educational Psychologist, 31*(3–4), 175–190.

Cobb, P., Yackel, E., & Wood, T. (1992). A constructivist alternative to the representational view in mathematics education. *Journal for Research in Mathematics Education, 23*(1), 2–33.

Cohen, E. (1997). *Athenian economy and society. A banking perspective.* Princeton, NJ: Princeton University Press.

Cozza Sayão, S. (2011). Entre o dizer e o dito: Sobre a precariedade e a finitude de nosso saber em Emmanuel Levinas [Between the saying and the said: On the precariousness and finitude of our knowledge in Emmanuel Levinas]. *Conjectura, 16*(1), 98–119.

Cremin, L. (1961). *The transformation of the school. Progressivism in American education, 1876–1957.* New York, NY: Vintage Books.

Critchney, S. (1999). *Ethics-politics-subjectivity: Essays on Derrida, Levinas, and contemporary French thought.* London, England: Verso.

Crump, T. (1990). *The anthropology of numbers.* Cambridge, England: Cambridge University Press.

Cuche, D. (2004). *La notion de culture dans les sciences sociales* [*The notion of culture in the social sciences*]. Paris, France: La Découverte.

D'Ambrosio, U. (1985). Ethnomathematics and its place in the history and pedagogy of mathematics. *For the learning of mathematics, 5*(1), 44–48.

D'Ambrosio, U. (2004). Una riflessione dell'etnomatematica: perché insegnare matematica? [A reflection on ethnomathematics: Why teach mathematics?]. In G. Arrigo (Ed.), *Atti del Convegno di didacttica della matematica* (pp. 29–37). Locarno, Switzerland.

D'Ambrosio, U. (2006). *Ethnomathematics.* Rotterdam, the Netherlands: Sense.

Damerow, P. (1996). *Abstraction and representation. Essays on the cultural evolution of thinking.* Dordrecht, the Netherlands: Kluwer.

D'Amore, B. (2015). Saber, conocer, labor en didáctica de la matemática: Una contribución a la teoría de la objetivación [To know, labour in didactics of mathematics: A contribution to the theory of objectivation]. In L. Branchetti (Ed.), *Teaching and learning mathematics. Some past and current approaches to mathematics education* (pp. 151–171). University of Urbino Carlo Bo, Italy: Isonomia. Retrieved from http://isonomia.uniurb.it/epistemologica

Daniels, H. (2002). *Vygotsky and pedagogy.* London, England: Routledge.

Darling, J., & Nordenbo, S. (2002). Progressivism. In N. Blake, P. Smeyers, R. Smith, & P. Standish (Eds.), *The philosophy of education* (pp. 288–308). Oxford, England: Blackwell.

Davydov, V. (1999). A new approach to the interpretation of activity structure and content. In S. Chaiklin, M. Hedegaard, & U. Jensen (Eds.), *Activity theory and social practice* (pp. 39–50). Aarhus, Denmark: Aarhus University Press.

Deacon, T. W. (1997). *The symbolic species.* New York, NY: W. W. Norton.

Dear, P. (1995). *Discipline and experience. The mathematical way in the scientific revolution.* Chicago, IL: The University of Chicago Press.

de Freitas, E., & Sinclair, N. (2013). New materialist ontologies in mathematics education: The body in/of mathematics. *Educational Studies in Mathematics, 83*, 453–470.

de Freitas, E., & Sinclair, N. (2014). *Mathematics and the body*. Cambridge, England: Cambridge University Press.

Deleuze, G. (1991). *Empiricism and subjectivity: An essay on Hume's theory of human nature*. New York, NY: Columbia University Press.

Del Río, P., & Álvarez, A. (1995). Directivity: The cultural and educational construction of morality and agency. Some questions arising from the legacy of Vygotsky. *Anthropology & Education Quarterly, 26*(4), 384–409.

Demosthenes. (2004). *Speeches 27–38* (D. M. MacDowell, Trans.). Austin, TX: University of Texas Press.

Demosthenes. (2011). *Speeches 39–49* (A. C. Scafuro, Trans.). Austin, TX: University of Texas Press.

Derry, J. (2017). An introduction to inferentialism in mathematics education. *Mathematics Education Research Journal, 29*, 403–418.

Descartes, R. (1886). *La géométrie* [*The geometry*]. Paris, France: A. Hermann, Librairie Scientifique.

Detienne, M. (1996). *The masters of truth in archaic Greece*. New York, NY: Zone Books.

Donham, D. L. (1999). *History, power, ideology: Central issues in Marxism and anthropology*. Berkeley, CA: University of California Press.

Dubbs, C. (2020). Whose ethics? Toward clarifying ethics in mathematics education research. *Journal of Philosophy of Education, 54*(3), 521–540.

Dupré, L. (1983). *Marx's social critique of culture*. New Haven, CT: Yale University Press.

Dzobo, N. K. (1980). The indigenous African theory of knowledge and truth. *The Conch, 7*(1–2), 85–102.

Eagleton, T. (1998). Body work. In S. Regan (Ed.), *The Eagleton reader* (pp. 157–162). Oxford, England: Blackwell.

Eagleton, T. (2000). *The idea of culture*. Oxford, England: Blackwell.

Eagleton, T. (2016). *Culture*. New Haven, CT: Yale University Press.

Edwards, S. (2009). Three versions of an ethics of care. *Nursing Philosophy, 10*, 231–240.

Elias, N. (2000). *The civilizing process: Sociogenetic and psycholgenetic investigations*. Oxford, England: Blackwell.

El'Konin, D. B. (1995). Problems in the psychology of activity. *Journal of Russian and East European Psychology, 33*(4), 32–34.

Ellerton, N. (1991). Classroom discourse and mathematics learning. In *Proceedings of the 15th conference of the international group for the Psychology of Mathematics Education* (Vol. 2, pp. 17–24). Assisi, Italy: PME.

Engeström, Y. (1999). Activity theory and individual and social transformation. In Y. Engeström, R. Miettinen, & R. L. Punamäki (Eds.), *Perspectives on activity theory* (pp. 19–38). Cambridge, England: Cambridge University Press.

Ernest, P. (1991). Constructivism, the psychology of learning, and the nature of mathematics: Some critical issues. In F. Furinghetti (Ed.), *Proceedings of 15th international conference on the Psychology of Mathematics Education* (Vol. 2, pp. 25–32). Assisi, Italy: PME.

Ernest, P. (2009). What is 'first philosophy' in mathematics education? In M. Tzekaki, M. Kaldrimidou, & H. Sakonidis (Eds.), *Proceedings of the 33rd conference of the international group for the Psychology of Mathematics Education* (Vol. 1, pp. 25–42). Thessaloniki, Greece: PME.

Farías Gutiérrez, M. (in press). La sensibilidad ética ofrecida a la educación [Ethical sensibility offered to education]. In L. Radford & M. Silva Acuña (Eds.), *Ética: Entre educación y filosofía*. Bogota, Colombia: Uniandes.

Farroni, T., & Menon, E. (2008). Visual perception and early brain development. In R. E. Tremblay, R. G. Barr, R. R. D. Peters, & M. Boivin (Eds.), *Encyclopedia on early childhood development* (pp. 1–6). Montreal, Québec: Centre of Excellence for Early Childhood Development. Retrieved May 5, 2009, from http://www.child-encyclopedia.com/documents/Farroni-MenonANGxp.pdf

Fay, B. (1996). *Contemporary philosophy of social science: A multicultural approach.* Oxford, England: Blackwell.

Fedoseyev, P. N., Oizerman, T., Melukhin, E., Ilyenkov, E., Lektorsky, V., et al. (1977). *Philosophy in the USSR – Dialectical materialism*. Moscow, Russia: Progress.

Fernandez-Armesto, F. (1998). *Truth: A history and a guide for the perplexed.* London, England: Black Swan Books.

Fernández Mouján, I. (2016). *Elogio de Paulo Freire*. Argentina: Noveduc.

Feuerbach, L. (1966). *Principles of the philosophy of the future.* Indianapolis, IN: Bobbs-Merrill.

Filloy, E., & Rojano, T. (1984). La aparición del lenguaje aritmético-algebraico [The emergence of the arithmetic-algebraic language]. *L'Educazione Matematica, 5*(3), 278–306.

Finley, M. (1973). *Studies in land and credit in ancient Athens*. New York, NY: Arno Press.

Fischbach, F. (2012). *Sans objet. Capitalisme, subjectivité, alienation* [*Without object. Capitalism, subjectivity, alienation*]. Paris, France: Vrin.

Fischbach, F. (2014). *La production des hommes* [*The production of men*]. Paris, France: Vrin.

Fischbach, F. (2015). *Philosophies de Marx* [*Marx's philosophies*]. Paris, France: Vrin.

Fortes, M. (1949). *The web of kinship among the Tallensi.* London, England: Oxford University Press.

Foucault, M. (1980). *Power/knowledge*. New York, NY: Pantheon Books.

Fowler, D. (1999). *The mathematics of Plato's academy*. Oxford, England: Clarendon Press.

Freire, P. (1970). *Pedagogia do oprimido*. Rio de Janeiro, Brazil: Paz e Terra.

Freire, P. (1998). *Pedagogy of freedom: Ethics, democracy, and civic courage*. Lanham, MD: Rowman & Littlefield.

Freire, P. (2004). *Pedagogy of indignation*. Boulder, CO: Paradigm.

Freire, P. (2016). *Pedagogia da solidariedade* [*Pedagogy of solidarity*]. São Paulo, Brazil: Paz & Terra.

Freitag, M. (2002). *L'oubli de la société: Pour une théorie critique de la postmodernité* [*The forgetting of society: For a critical theory of postmodernity*]. Québec, Canada: Les Presses de l'Université Laval.

Fried, M. (2009). Similarity and equality in Greek mathematics: Semiotics, history of mathematics, and mathematics education. *For the Learning of Mathematics, 29*(1), 2–7.

Gal'perin, P. (1967). On the notion of internalization. *Soviet Psychology, 5*(3), 28–33.

Gal'perin, P. (1989). Organization of mental activity and the effectiveness of learning. *Soviet Psychology, 27*(3), 65–82.

Gal'perin, P. (1992). Stage-by-stage formation as a method of psychological investigation. *Journal of Russian & East European Psychology, 30*(4), 60–80.

Gardner, H. (1970). Piaget and Lévi-Strauss: The quest for mind. *Social Research, 37*(3), 348–365.

Geertz, C. (1973). *The interpretation of cultures*. New York, NY: Basic Books.

Geurts, K. L. (2002). *Culture and the senses*. Berkeley, CA: University of California Press.

Giroux, H. (1986). Authority, intellectuals, and the politics of practical learning. *Teachers College Record, 88*(1), 22–40.

Giroux, H. (2005). *Border crossings*. London, England: Routledge.

Godelier, M. (2010). *L'idéel et le matériel. Pensée, économie, sociétés* [*The ideal and the material. Thought, economy, societies*]. Paris, France: Champs Essais.

Godino, J., & Burgos, M. (2020). ¿Cómo enseñar las matemáticas y ciencias experimentales? Resolviendo el dilema entre transmisión e indagación [How to teach math and experimental science? Solving the dilemma between transmission and inquiry]. *Revista Paradigma, 41*, 80–106.

González Rey, F. (2009). Historical relevance of Vygotsky's work: Its significance for a new approach to the problem of subjectivity in psychology. *Outlines, 1*, 59–73.

González Rey, F. (2011a). The path to subjectivity. In P. Portes & S. Salas (Eds.), *Vygotsky in 21st century society* (pp. 32–49). New York, NY: Peter Lang.

González Rey, F. (2011b). A re-examination of defining moments in Vygotsky's work and their implications for his continuing legacy. *Mind, Culture, and Activity, 18*, 257–275.

Gow, J. (2010). *A short history of Greek mathematics*. Cambridge, England: Cambridge University Press.

Grootenboer, P., & Jorgensen (Zevenbergen), R. (2009). Towards a theory of identity and agency in coming to learn mathematics. *Eurasia Journal of Mathematics, Science & Technology Education, 5*(3), 255–266.

Guilherme, M. (2017). Freire's philosophical contribution for a theory of intercultural ethics: A deductive analysis of his work. *Journal of Moral Education, 46*(4), 422–434.

Guillemette, D. (2017). History of mathematics in secondary school teachers' training: Towards a nonviolent mathematics education. *Educational Studies in Mathematics, 96*, 349–365.

Harré, R. (2012). Positioning theory: Moral dimensions of social-cultural psychology. In J. Valsiner (Ed.), *The Oxford handbook of culture and psychology* (pp. 191–206). New York, NY: Oxford University Press.

Harré, R., & Gillett, G. (1994). *The discursive mind*. London, England: Sage.

Harris, P. (1991). *Mathematics in a cultural context: Aboriginal perspectives on space, time and money*. Greelong, Australia: Deakin University.

Hatle, J. D., & Salazar, B. A. (2001). Aposematic coloration of gregarious insects can delay predation by ambush predator. *Environmental Entomology, 30*(1), 51–54.

Heath, L. T. (1908). *The thirteen books of the elements* (Vol. 1, Dover edition, T. L. Heath, Trans.). Cambridge, England: Cambridge University Press.

Hegel, G. (1991). *The encyclopaedia logic. Part I of the encyclopaedia of philosophical sciences* (T. F. Geraets, W. A. Suchting, & H. S. Harris, Trans.). Indianapolis, IN: Hackett.

Hegel, G. (1994). *Concept préliminaire de l'encyclopédie des sciences philosophiques en agrégé* [*Short preliminary concept of the encyclopedia of philosophical sciences*] (B. Bourgeois, Trans.). Paris, France: Vrin.

Hegel, G. (2001). *The philosophy of history*. Kitchener, ON: Batoche Books. (Original work published 1837)

Hegel, G. (2009). *Logic* (W. Wallace, Trans.). Pacifica, CA: MIA. (Original work published 1830)

Heidegger, M. (1977). *The question concerning technology and other essays*. New York, NY: Harper Torchbooks.

Heidegger, M. (2004). *On the essence of language*. New York, NY: SUNY. (Original work published 1999)

Henry, M. (1976). *Marx*. Paris, France: Gallimard.

Herbel-Eisenmann, B. A., Wagner, D., Johnson, K., Suh, H., & Figueras, H. (2015). Positioning in mathematics education: Revelations on an imported theory. *Educational Studies in Mathematics, 89*, 185–204.

Herodotus. (2004). *The histories* (G. C. Macaulay, Trans.). New York, NY: Barnes & Noble. Epub version.

Hirata, S., Morimura, N., & Houki, C. (2009). How to crack nuts: Acquisition process in captive chimpanzees (pan troglodytes) observing a model. *Animal Cognition, 12*, 87–101. doi:10.1007/s10071-009-0275-3

Hobbes, T. (1841). *The English works of Thomas Hobbes of Malmesbury* (Vol. 2, W. Molesworth, Ed.). London, England: John Bohn.

hooks, b. (2015). *Thinking feminist, thinking Black*. London, England: Routledge.

Horkheimer, M., & Adorno, T. W. (2002). *Dialectic of enlightenment: Philosophical fragments*. Stanford, CA: Stanford University Press.

Høyrup, J. (2002). *Lengths, widths, surfaces. A portrait of old Babylonian algebra and its kin*. New York, NY: Springer.

Høyrup, J. (2008). The tortuous ways toward a new understanding of algebra in the Italian abbacus school (14th–16th centuries). In O. Figueras, J. L. Cortina, S. Alatorre, T. Rojano, & A. Sepúlveda (Eds.), *Proceedings of the joint 32nd conference of the international group for the Psychology of Mathematics Education and the 30th North American chapter* (Vol. 1, pp. 1–15). Morelia, Mexico.

Hughes, B. (1986). Gerard of Cremona's translation of Al-Khwarizmi's Al-Jabr: A critical edition. *Mediaeval Studies, 48*, 211–263.

Husserl, E. (1931). *Ideas. General introduction to pure phenomenology* (W. R. Boyce Gibson, Trans., 3rd [1958] ed.). London, England: George Allen & Unwin.

Husserl, E. (1970). *Logical investigations*. London, England: Routledge and K. Paul.

Ilyenkov, E. V. (1977). *Dialectical logic*. Moscow, Russia: Progress.

Inwood, M. (1992). *A Hegel dictionary*. Oxford, England: Blackwell.

Itard, J. M. G. (1962). *The wild boy of Aveyron*. New York, NY: Meredith.

Jackson, L. (2007). The individualist? The autonomy of reason in Kant's philosophy and educational views. *Studies in Philosophy of Education, 26*, 335–344.

Jacotot, J. (1828). *Enseignement universel, mathématiques* [*Universal teaching, mathematics*]. Louvain, Belgium: G. Guelens.

Jameson, M. H. (1977). Agriculture and slavery in classical Athens. *The Classical Journal, 73*(2), 122–145.

Janet, P. (1929). *L'evolution psychologique de la personnalité* [*The psychological evolution of personality*]. Paris, France: Masson.

Janßen, T., & Radford, L. (2015). Solving equations: Gestures, (un)allowable hints, and the unsayable matter. In *Proceedings of the 9th Congress of the European Society for Research in Mathematics Education* (CERME) (pp. 419–425). Prague, Czechia: Charles University.

Jasim, S. A., & Oates, J. (1986). Early tokens and tablets in Mesopotamia: New information from Tell Abada and Tell Brak. *World Archeology, 17*(3), 348–362.

Juzwik, M. (2004). Towards an ethics of answerability: Reconsidering dialogism in sociocultural literacy research. *College Composition and Communication, 55*(3), 536–567.

Kant, I. (2003). *Critique of pure reason* (N. K. Smith, Trans.). New York, NY: Palgrave Macmillan. (Original work published 1781)

Kant, I. (2006). *Groundwork of the metaphysic of morals* (M. Gregor, Trans.). Cambridge, England: Cambridge University Press.

Kant, I. (2010). *Political writings* (H. Reiss, Ed.). New York, NY: Cambridge University Press.

Kawagley, O. (1990). Yup'ik ways of knowing. *Canadian Journal of Native Education, 17*, 5–17.

Kieran, C. (2018). *Teaching and learning algebraic thinking with 5- to 12-year-olds: The global evolution of an emerging field of research and practice*. New York, NY: Springer.

Kozulin, A., Gindis, B., Ageyev, V., & Miller, S. (2005). *Vygotsky's educational theory in cultural context*. Cambridge, England: Cambridge University Press.

Kuzniak, A., Tanguay, D., & Elia, I. (2016). Mathematical working spaces in schooling: An introduction. *ZDM, 48*, 721–737.

LaCroix, L. (2014). Learning to see pipes mathematically: Preapprentices' mathematical activity in pipe trades training. *Educational Studies in Mathematics, 86*, 157–176.

Lancy, D. F. (1983). *Cross-cultural studies in cognition and mathematics*. New York, NY: Academic Press.

Lang, M. (1957). Herodotos and the abacus. *Hesperia, 26*(3), 271–288.

Lang, M. (1964). The abacus and the calendar. *Hesperia, 33*(2), 146–167.

Lang, M. (1968). Abaci from the Athenian agora. *Hesperia, 37*(2), 241–243.

Lasprilla Herrera, A., Radford, L., & León, O. (in press). Formas de interacción social y aspectos éticos en actividades matemáticas escolares [Forms of social interaction and ethical aspects in school mathematics activities]. In L. Radford & M. Silva Acuña (Eds.), *Ética: Entre educación y filosofía*. Bogota, Colombia: Uniandes.

Lave, J. (1988). *Cognition in practice*. Cambridge, England: Cambridge University Press.

Lave, J., & Wenger, E. (1991). *Situated learning: Legitimate peripheral participation*. Cambridge, England: Cambridge University Press.

Lawlor, R., & Lawlor, D. (1979). *Theon of Smirna: Mathematics useful for understanding Plato*. San Diego, CA: Wizards Bookshelf.

Lea, H. (1990). Spatial concepts in the Kalahari. In T. Booker, P. Cobb, & G. Mendicuti (Eds.), *Proceedings of the 14th conference of the international group for the Psychology of Mathematics Education* (Vol. 2, pp. 259–266). Mexico DF: PME.

Lefebvre, H. (2009). *Dialectical materialism*. Minneapolis, MN: University of Minnesota Press.

Lefèvre, W. (1981). Rechenstein und sprache [Computing stones and language]. In P. Damerow & W. Lefèvre (Eds.), *Rechenstein, experiment, sprache. Historische fallstudien zur entstehung der exakten wissenschaften* (pp. 115–169). Stuttgart, Germany: Klett-Cotta.

Leontiev [or Leont'ev], A. N. (1968). *El hombre y la cultura. Problemas téoricos sobre educación* [*Man and culture. Theoretical problems on education*]. Mexico: Editorial Grijalbo.

Leont'ev, A. N. (1974). The problem of activity in psychology. *Soviet Psychology, 13*(2), 4–33.

Leont'ev, A. N. (1978). *Activity, consciousness, and personality*. Englewood Cliffs, NJ: Prentice-Hall.

Leont'ev, A. N. (2005). The genesis of activity. *Journal of Russian and East European Psychology, 43*(4), 58–71.

Leontyev [or Leont'ev], A. N. (2009a). *Activity and consciousness.* Pacifica, CA: MIA. Retrieved August 29, 2009, from http://www.marxists.org/archive/leontev/works/activity-consciousness.pdf

Leontyev, A. N. (2009b). *The development of mind.* Pacifica, CA: MIA.

Lerman, S. (1992). The function of language in radical constructivism: A Vygotskian perspective. In W. Geeslin & K. Graham (Eds.), *Proceedings of the 16th conference of the international group for the Psychology of Mathematics Education* (Vol. 2, pp. 40–47). Durham, NH: PME.

Lerman, S. (1996a). Guest editorial. *Educational Studies in Mathematics* (Special Issue: *Socio-cultural Approaches to Mathematics Teaching and Learning*), *31*(1–2), 1–9.

Lerman, S. (1996b). Intersubjectivity in mathematics learning: A challenge to the radical constructivist paradigm? *Journal for Research in Mathematics Education, 27*(2), 133–150.

Lerman, S. (2006). Socio-cultural research in PME. In A. Gutiérrez & P. Boero (Eds.), *Handbook of research on the Psychology of Mathematics Education: Past, present and future* (pp. 347–366). Rotterdam, the Netherlands: Sense.

Lettvin, J. Y., Maturana, H. R., McCulloch, W. S., & Pitts, W. H. (1959). What the frog's eye tells the frog's brain. *Proceedings of the Institute of Radio Engineers, 47*(11), 1940–1951.

Levant, A., & Oittinen, V. (2014). *Dialectics of the ideal.* Leiden, the Netherlands: Brill.

Levi, A. W. (1974). *Philosophy as social expression.* Chicago, IL: The University of Chicago Press.

Lévinas, E. (1978). *Autrement qu'être ou au-delà de l'essence* [*Otherwise than being or beyond essence*]. The Hague, the Netherlands: Martinus Nijhoff.

Levinas, E. (1989). *The Levinas reader* (S. Hand, Ed.). Oxford, England: Blackwell.

Lévinas, E. (1982). *Éthique et infini* [*Ethics and infinity*]. Paris, France: Fayard.

Levinas, E. (2006). *Otherwise than being or beyond essence.* Pittsburgh, PA: Duquesne University Press.

Lévi-Strauss, C. (1962). *La pensée sauvage* [*The savage mind*]. Paris, France: Plon.

Lévy-Bruhl, L. (1922). *La mentalité primitive.* Paris, France: Presses Universitaires de France.

Lévy-Bruhl, L. (1928). *Les fonctions mentales dans les sociétés inférieures* (9th ed.). Paris, France: Librairie Félix Alcan.

Lévy Zumwalt, R. (2019). *Franz Boas.* Lincoln, NE: University of Nebraska Press.

Ligozat, F., & Schubauer-Leoni, M.-L. (2010). The joint action theory in didactics: Why do we need it in the case of teaching and learning mathematics? In V. Durand-Guerrier, S. Soury-Lavergne, & F. Arzarello (Eds.), *Proceedings of the 6th*

Conference of European Research in Mathematics Education (CERME 6) (pp. 1615–1624). Lyon, France: Institut National de la Recherche Pédagogique.

Lizcano, E. (2009). *Imaginario colectivo y creación matemática* [*Collective imagination and mathematical creation*]. Madrid, Spain: Gedisa.

Locke, J. (1825). *An essay concerning human understanding*. London, England: Thomas Davison. (Original work published 1690)

Løgstrup, K. (1997). *The ethical demand*. Notre Dame, IN: University of Notre Dame Press.

Luria, A. R. (1931). Psychological expedition to Central Asia. *Science, 74*, 383–384.

Luria, A. R. (1934). The second psychological expedition to Central Asia. *Journal of Genetic Epistemology, 41*, 255–259.

Luria, A. R. (1979). *The making of mind*. Cambridge, MA: Harvard University Press.

Macherey, P. (2008). *Marx 1845. Les thèses sur Feuerbach* [*Marx 1845. The theses on Feuerbach*]. Paris, France: Éditions Amsterdam.

MacIntyre, A. (1966). *A short history of ethics*. New York, NY: Touchstone.

MacIntyre, A. (2007). *After virtue*. Notre Dame, IN: University of Notre Dame Press.

Maheux, J.-F., & Proulx, J. (2017). Éthique et activité mathématique [Ethics and mathematical activity]. *Éducation et francophonie, 45*(1), 174–194.

Maheux, J.-F., & Roth, M.-W. (2014). The relationality in/of teacher-student communication. *Mathematics Education Research Journal, 26*, 503–529.

Malafouris, L. (2004). The cognitive basis of material engagement: Where brain, body and culture conflate. In E. DeMarrais, C. Gosden, & C. Renfrew (Eds.), *Rethinking materiality: The engagement of mind with the material world* (pp. 53–61). Cambridge, England: McDonald Institute Monographs.

Malafouris, L., & Renfrew, C. (2010). The cognitive life of things: Archaeology, material engagement and the extended mind. In L. Malafouris & C. Renfrew (Eds.), *The cognitive life of things: Recasting the boundaries of the mind* (pp. 1–12). Cambridge, England: McDonald Institute Monographs.

Manitius, K. (1888). *Des Hypsikles Schrift Anaphorikos nach Überlieferung und Inhalt kritisch behandelt* [*Hypsikles' text "anaphorikos" according to transmission and content dealt with in a critical way*]. Dresden, Germany: Lehmannsche Buchdruckerei.

Markus, G. (1986). *Language and production. A critique of paradigms*. Dordrecht, the Netherlands: Reidel.

Martin, J. (2004). The educational inadequacy of conceptions of self in educational psychology. *Interchange: A Quarterly Review of Education, 35*, 185–208.

Marx, A., & Burnham, S. (1997). *Musical form in the age of Beethoven*. Cambridge, England: Cambridge University Press.

Marx, K. (1904). *A contribution to the critique of political economy* (N. Stone, Trans.). Chicago, IL: Charles H. Kerr.

Marx, K. (1968). *Oeuvres. Économie II [Works. Economy, Vol. 2]* (M. Rubel, Ed.). Paris, France: Pléiade.

Marx, K. (1973). *Grundrisse: Introduction to the critique of political economy*. Baltimore, MD: Penguin Books.

Marx, K. (1988). *Economic and philosophic manuscripts of 1844*. Amherst, NY: Prometheus Books. (Original work published 1932)

Marx, K. (1998). *The German ideology*. New York, NY: Prometheus Books.

Marx, K., & Engels, F. (1956). *The holy family or critique of critical critique*. Moscow, Russia: Foreign Languages Publishing House.

Masi, M. (1983). *Boethian number theory. A translation of the De institutione arithmetica*. Amsterdam, the Netherlands: Éritions Rodopi.

Matsuzawa, T., Biro, D., Humle, T., Inoue-Nakamura, N., Tonooka, R., & Yamakoshi, G. (2001). Emergence of culture in wild chimpanzees: Education by master-apprenticeship. In T. Matsuzawa (Ed.), *Primate origins of human cognition and behavior* (pp. 557–574). Tokyo, Japan: Springer.

Matusov, E. (1998). When solo activity is not privileged: Participation and internalization models of development. *Human Development, 41*, 326–349.

Maybee, J. (2009). *Picturing Hegel*. Lanham, MD: Lexington Books.

Meira, L. (1998). Making sense of instructional devices: The emergence of transparency in mathematical activity. *Journal for Research in Mathematics Education, 29*(2), 121–142.

Meritt, B. D. (1946). Borrowings in the Archidamian war. *The Classical Quarterly, 40*(1–2), 60–64.

Mészáros, I. (2010). *Social structure and forms of consciousness*. New York, NY: Monthly Review Press.

Mikhailov, F. T. (1980). *The riddle of the self*. Moscow, Russia: Progress.

Min, A. (1998). Toward a dialectic of totality and infinity: Reflections on Emmanuel Levinas. *The Journal of Religion, 78*(4), 571–592.

Ministère de l'éducation de l'Ontario. (2005). *Le curriculum de l'Ontario de la 1er à la 8e année. Mathématiques [The Ontario curriculum of mathematics, Grades 1–8]*. Ottawa, Ontario: Imprimeur de la Reine pour l'Ontario.

Miranda, I., Radford, L., & Guzmán, J. (2007). Interpretación de gráficas cartesianas sobre el movimiento desde el punto de vista de la teoría de la objetivación. *Educación Matemática, 19*(3), 1–26.

Moretti, V., Panossian, M. L., & Radford, L. (2018). Questões em torno da teoria da objetivação. *Obutchénie, 2*(1), 230–251.

Morgan, C. (2006). What does social semiotics have to offer mathematics education research? *Educational Studies in Mathematics, 61*, 219–245.

Morgan, C. (2018). Using social semiotics to explore institutional assumptions about mathematics students and teachers. In N. Presmeg, L. Radford, M. Roth, & G. Kadunz (Eds.), *Signs of signification: Semiotics in mathematics education research* (pp. 139–154). Cham, Switzerland: Springer.

Nafissi, M. (2004). Class, embeddedness, and the modernity of ancient Athens. *Comparative Studies in Society and History, 46*(2), 378–410.

Nealon, J. (1997). The ethics of dialogue: Bakhtin and Levinas. *College English, 59*(2), 129–148.

Nealon, J. (1998). *Alterity politics. Ethics and performative subjectivity.* Durham, NC: Duke University Press.

Netz, R. (1998). Greek mathematical diagrams: Their use and their meaning. *For the Learning of Mathematics, 18*(3), 33–39.

Netz, R. (1999). *The shaping of deduction in Greek mathematics.* Cambridge, England: Cambridge University Press.

Netz, R. (2002). Counter culture: Towards a history of Greek numeracy. *History of Sciences, 40*(3), 321–352.

Nicomachus of Geresa. (1924). *Introduction to arithmetic* (D'ooge, Trans.; with studies in Greek arithmetic by F. E. Robbins & L. C. Karpinski). London, England: Macmillan.

Niss, M. (1999). Aspects of the nature and state of research in mathematics education. *Educational Studies in Mathematics, 40*, 1–24.

Nodari, P. C. (2010). *Sobre ética* [*On ethics*]. Caxias do Sul, Rio Grande do Sul: Educs.

Novotný, K. (2016). *Corps et affectivité dans Autrement qu'être ou au-delà de l'essence* [*Body and affectivity in Otherwise than being or beyond essence*]. In D. Cohen-Levinas & A. Schnell (Eds.), *Relire Autrement qu'être ou au-delà de l'essence d'Emmanuel Levinas* (pp. 121–133). Paris, France: Vrin.

Ontario Ministry of Education. (2014). *Achieving excellence. A renewed vision for education in Ontario.* Ottawa, Ontario: Queen's Printer for Ontario.

Otte, M. (1998). Limits of constructivism: Kant, Piaget and Peirce. *Science & Education, 7*, 425–450.

Owens, K. (2001). Indigenous mathematics: A rich diversity. Mathematics: Shaping Australia. *Proceedings of the 18th biennial conference of the Australian Association of Mathematics Teachers, 151–167.* Retrieved February 17, 2006, from http://www.aamt.edu.au/ICSIMAN/resources/papers/owens.pdf

Pais, A. (2013). An ideology critique of the use-value of mathematics. *Educational Studies in Mathematics, 84*, 15–34.

Pais, A., & Valero, P. (2012). Researching research: Mathematics education in the political. *Educational Studies in Mathematics, 80*, 9–24.

Peirce, C. S. (1931–1958). *Collected papers* (CP, Vols. 1–8). Cambridge, MA: Harvard University Press.

Piaget, J. (1970). *Genetic epistemology*. New York, NY: W. W. Norton.

Piaget, J., & Garcia, R. (1989). *Psychogenesis and the history of science*. New York, NY: Columbia University Press.

Pinar, W. F., & Irwin, R. L. (2005). *Curriculum in a new key: The collected works of Ted T. Aoki*. Mahwah, NJ: Lawrence.

Plato. (1961). *The collected dialogues of Plato including the letters* (E. Hamilton & H. Cairns, Eds.). New York, NY: Pantheon Books.

Poovey, M. (1998). *A history of the modern fact: Problems of knowledge in the sciences of wealth and society*. Chicago, IL: University of Chicago Press.

Popkewitz, T. (2004). The alchemy of the mathematics curriculum: Inscriptions and the fabrication of the child. *American Educational Research Journal, 41*(1), 3–34.

Popkewitz, T. (2009). Globalization as a system of reason: The historical possibility and the political in pedagogical policy and research. In T. Popkewitz & F. Rizvi (Eds.), *Globalization and the study of education* (pp. 247–267). Hoboken, NJ: Wiley-Blackwell.

Presmeg, N., Radford, L., Roth, W., & Kadunz, G. (2016). *Semiotics in mathematics education*. Cham, Switzerland: Springer.

Pumuge, H. M. (1975). The counting system of the Pekai-Alue tribe of the Topopul village in the Ialibu sub-district in the Southern Highlands district, Papua New Guinea. *Science in New Guinea, 3*(1), 19–25.

Radford, L. (1995). L'émergence et le développement conceptuel de l'algèbre [The emergence and conceptual development of algebra]. In F. Lalonde, J. Jaboeuf, & Y. Nouazé (Eds.), *Proceedings of the 1st European summer university "history and epistemology in mathematics education"* (pp. 69–83). Montpellier, France: IREM de Montpellier.

Radford, L. (2000). Signs and meanings in students' emergent algebraic thinking: A semiotic analysis. *Educational Studies in Mathematics, 42*(3), 237–268.

Radford, L. (2004). Cose sensibili, essenze, oggetti matematici ed altre ambiguità [Sensible things, essences, mathematical objects and other ambiguities]. *La Matematica e la sua Didattica, 1*, 4–23. (English version retrieved from http://luisradford.ca)

Radford, L. (2005). The semiotics of the schema. Kant, Piaget, and the calculator. In M. Hoffmann, J. Lenhard, & F. Seeger (Eds.), *Activity and sign. Grounding mathematics education* (pp. 137–152). New York, NY: Springer.

Radford, L. (2006). The cultural-epistomological conditions of the emergence of algebraic symbolism. In F. Furinghetti, S. Kaijser, & C. Tzanakis (Eds.), *Proceedings of the 2004 conference of the international study group on the relations between the history and pedagogy of mathematics & ESU 4* (rev. ed., pp. 509–524). Uppsala, Sweden.

Radford, L. (2007). Towards a cultural theory of learning. In D. Pitta-Pantazi & G. Philippou (Eds.), *Proceedings of the fifth Congress of the European Society for Research in Mathematics Education* (CERME-5) (pp. 1782–1797). Lamaca, Cyprus.

Radford, L. (2008a). Connecting theories in mathematics education: Challenges and possibilities. *ZDM, 40*(2), 317–327.

Radford, L. (2008b). Iconicity and contraction: A semiotic investigation of forms of algebraic generalizations of patterns in different contexts. *ZDM, 40*(1), 83–96.

Radford, L. (2008c). Culture and cognition: Towards an anthropology of mathematical thinking. In L. English (Ed.), *Handbook of international research in mathematics education* (2nd ed., pp. 439–464). New York, NY: Routledge, Taylor and Francis.

Radford, L. (2010). Algebraic thinking from a cultural semiotic perspective. *Research in Mathematics Education, 12*(1), 1–19.

Radford, L. (2011). Grade 2 students' non-symbolic algebraic thinking. In J. Cai & E. Knuth (Eds.), *Early algebraization* (pp. 303–322). Berlin, Germany: Springer-Verlag.

Radford, L. (2012a). On the development of early algebraic thinking. *PNA, 6*(4), 117–133.

Radford, L. (2012b). Education and the illusions of emancipation. *Educational Studies in Mathematics, 80*(1), 101–118.

Radford, L. (2013a). Sensuous cognition. In D. Martinovic, V. Freiman, & Z. Karadag (Eds.), *Visual mathematics and cyberlearning* (pp. 141–162). New York, NY: Springer.

Radford, L. (2013b, November 6–8). Sumisión, alienación y (un poco de) esperanza: hacia una visión cultural, histórica, ética y política de la enseñanza de las matemáticas [Submission, alienation and (a little of) hope: Towards a cultural, historical, ethical and political vision of the teaching of mathematics. In A. Ramirez & Y. Morales (Eds.), *Memorias del I congreso de educación matemática de América Central y El Caribe*. Santo Domingo, República Dominicana. Retrieved from http://luisradford.ca

Radford, L. (2014a). Towards an embodied, cultural, and material conception of mathematics cognition. *ZDM, 46*, 349–361.

Radford, L. (2014b). On teachers and students. In P. Liljedahl, C. Nicol, S. Oesterle, & D. Allan (Eds.), *Proceedings of the joint 38th conference of the international group for the Psychology of Mathematics Education and the 36th conference of the American chapter* (Vol. 1, pp. 1–20). Vancouver, BC: PME.

Radford, L. (2014c). The progressive development of early embodied algebraic thinking. *Mathematics Education Research Journal, 26*(2), 257–277.

Radford, L. (2016a). The epistemic, the cognitive, the human: A commentary on the mathematical working space approach. *ZDM, 48*, 925–933.

Radford, L. (2016b). On alienation in the mathematics classroom. *International Journal of Educational Research, 79*, 258–266.

Radford, L. (2017a). Mathematics education theories: The question of their growth, connectivity, and affinity. *La Matematica e la sua Didattica, 25*(2), 217–228.

Radford, L. (2017b). On inferentialism. *Mathematics Education Research Journal, 29*, 493–508.

Radford, L. (2018a). On theories in mathematics education and their conceptual differences. In B. Sirakov, P. de Souza, & M. Viana (Eds.), *Proceedings of the international congress of mathematicians* (Vol. 4, pp. 4055–4074). Singapore: World Scientific Publishing.

Radford, L. (2018b). Algunos desafíos encontrados en la elaboración de la teoría de la objetivación [Some challenges encountered in the development of the theory of objectification]. *PNA, 12*(2), 61–80.

Radford, L. (2018c). The emergence of symbolic algebraic thinking in primary school. In C. Kieran (Ed.), *Teaching and learning algebraic thinking with 5- to 12-year-olds: The global evolution of an emerging field of research and practice* (pp. 3–25). New York, NY: Springer.

Radford, L. (2018d). A plea for a critical transformative philosophy of mathematics education. In P. Ernest (Ed.), *The philosophy of mathematics education today* (pp. 1–10). Cham, Switzerland: Springer.

Radford, L. (2019). So, you say that doing math is like playing music? The mathematics classroom as a concert hall. *La matematica e la sua didattica, 27*(1), 69–87.

Radford, L. (2020a). L'ethnomathématique au carrefour de la recolonisation et la décolonisation des savoirs [Ethnomathematics at the crossroads of recolonization and decolonization of knowledge]. In G. Maheux, S. Quintriqueo, G. Pellerin, & L. Bacon (Eds.), *La décolonisation de la scolarisation des jeunes inuit et des Première Nations: sens et défis* (pp. 247–276). Montréal, QC: Presses de l'Université du Québec.

Radford, L. (2020b). Le concept de travail conjoint dans la théorie de l'objectivation. In M. Flores González, A. Kuzniak, A. Nechache, & L. Vivier (Eds.), *Cahiers du laboratoire de didactique André Revuz n°21* (pp. 19–41). Paris, France: IREM de Paris.

Radford, L. (2020c). Davydov's concept of the concept and its dialectical materialist background. *Educational studies in Mathematics* [Online first]. https://doi.org/10.1007/s10649-020-09959-y

Radford, L. (2020d). Play and the production of subjectivities in preschool. In M. Carlsen, I. Erfjord, & P. S. Hundeland (Eds.), *Mathematics education in the early years. Results from the POEM4 conference 2018* (pp. 43–60). Cham, Switzerland: Springer.

Radford, L. (2020e). Les finalités éducatives scolaires en mathématiques: présupposés, égarements et quelques pistes pour retrouver la voie [The educational aims of school mathematics: presuppositions, diversions and some ways to find the way back]. In Y. Lenoir (Ed.), *Les finalités éducatives scolaires. Une étude critique des approches théoriques, philosophiques et idéologiques* (Vol. 2, pp. 353–371). Saint-Lambert, QC: Éditions Cursus Universitaire.

Radford, L. (2020f). ¿Cómo sería una actividad de enseñanza-aprendizaje que busca ser emancipadora? La labor conjunta en la teoría de la objetivación [What would an emancipatory teaching-learning activity look like? Joint labour in the theory of objectivation]. *Revista Colombiana de Matemática Educativa, RECME, Número especial de la Teoría de la Objetivación, 5*(2), 15–31.

Radford, L. (forthcoming). Sensed objects, sensing subjects: Embodiment from a dialectical materialist perspective. In L. Edwards & C. Krause (Eds.), *The body in mathematics*. Leiden, the Netherlands: Brill | Sense.

Radford, L. (in press). Corps, matière et signes dans la constitution du sens en mathématiques [Body, matter and signs in the constitution of meaning in mathematics]. In C. Houdement, C. Hache, & C. de Hosson (Eds.), *Sémiotique et apprentissages scientifiques*. Paris, France: ISTE Editions.

Radford, L., Arzarello, F., Edwards, L., & Sabena, C. (2017). The multimodal material mind: Embodiment in mathematics education. In J. Cai (Ed.), *First compendium for research in mathematics education* (pp. 700–721). Reston, VA: NCTM.

Radford, L., & Barwell, R. (2016). Language in mathematics education research. In A. Gutiérrez, G. Leder, & P. Boero (Eds.), *The second handbook of research on the psychology of Mathematics Education. The journey continues* (pp. 275–313). Rotterdam, the Netherlands: Sense.

Radford, L., & Demers, S. (2004). *Communication et apprentissage. Repères conceptuels et pratiques pour la salle de classe de mathématiques* [Communication and learning. Conceptual and practical benchmarks for the mathematics classroom]. Ottawa, ON: Centre franco-ontarien des ressources pédagogiques.

Radford, L., Demers, S., Guzmán, J., & Cerulli, M. (2003). Calculators, graphs, gestures, and the production meaning. In P. Pateman, B. Dougherty, & J. Zilliox (Eds.), *Proceedings of the 27th conference of the international group for the Psychology of Mathematics Education* (PME27-PMENA25) (Vol. 4, pp. 55–62). Honolulu, HI: University of Hawaii.

Radford, L., Demers, S., Guzmán, J., & Cerulli, M. (2004). The sensual and the conceptual: Artefact-mediated kinesthetic actions and semiotic activity. In J. Høines & A. Fuglestad (Eds.), *Proceedings of the 28th conference of the international group for the Psychology of Mathematics Education* (PME28). (Vol. 4, pp. 73–80). Bergen University College.

Radford, L., Demers, S., & Miranda, I. (2009). *Processus d'abstraction en mathématiques* [*Processes of abstraction in mathematics*]. Ottawa, Ontario: Centre franco-ontarien de ressources pédagogiques, Imprimeur de la Reine pour l'Ontario.

Radford, L., & Grenier, M. (1996). Entre les idées, les choses et les symboles. Une séquence d'enseignement d'introduction à l'algèbre. *Revue des sciences de l'éducation, 22*, 253–276.

Radford, L., & Lasprilla Herrera, A. (2020). De porqué la ética es ineludible de considerar en la enseñanza-aprendizaje de las matemáticas [Why ethics is a unavoidable in the teaching-learning of mathematics]. *La matematica e la sua didattica, 28*(1), 107–128.

Radford, L., & Roth, W. (2017). Alienation in mathematics education: A problem considered from neo-Vygotskian approaches. *Educational Studies in Mathematics, 96*(3), 367–380.

Radford, L., & Sabena, C. (2015). The question of method in a Vygotskian semiotic approach. In A. Bikner-Ahsbahs, C. Knipping, & N. Presmeg (Eds.), *Approaches to qualitative research in mathematics education* (pp. 157–182). New York, NY: Springer.

Ratner, C. (2000). Agency and culture. *Journal of the Theory of Social Behaviour, 30*, 413–434.

Ratner, C. (2015). Classic and revisionist sociocultural theory, and their analyses of expressive language: An empirical and theoretical assessment. *Language and Sociocultural Theory, 2*(1), 51–83.

Resnick, L., & Wolff, R. (1982). Marxist epistemology: The critique of economic determinism. *Social Text, 6*, 31–72.

Resnick, S., & Wolff, R. (2011). Persistence du Marxisme traditionnel et pertinence du marxisme surdéterministe [The persistence of traditional Marxism and the relevance of overdeterministic Marxism]. *Actuel Marx, 50*, 136–152.

Ricard, M.-A. (1999). La dialectique de T. W. Adorno [The dialectics of T. W. Adorno]. *Laval Théologique et Philosophique, 55*(2), 267–283.

Rivera, F. (2011). *Toward a visually-oriented school mathematics curriculum.* Dordrecht, the Netherlands: Springer.

Robbins, F. E. (1920). Posidonius and the sources of Pythagorean arithmology. *Classical Philology, 15*(4), 309–322.

Robbins, F. E. (1921). The tradition of Greek arithmology. *Classical Philology, 16*(2), 97–123.

Roberts, K. (2011). *The origin of business, money, and markets.* New York, NY: Columbia University Press.

Rogoff, B. (1990). *Apprenticeship in thinking.* Oxford, England: Oxford University Press.

Röhrs, H., & Lenhart, V. (1995). *Progressive education across the continents.* Frankfurt and Main, Germany: Peter Lang.

Rojano, T. (1996). The role of problems and problem solving in the development of algebra. In N. Bednarz, C. Kieran, & L. Lee (Eds.), *Approaches to algebra: Perspectives for research and teaching* (pp. 55–62). Dordrecht, the Netherlands: Kluwer.

Roth, G. (1986). Neural mechanisms of prey recognition: An example in amphibians. In M. Feder & G. Lauder (Eds.), *Predator-prey relationships: Perspectives and approaches from the study of lower vertebrates* (pp. 42–68). Chicago, IL: The University of Chicago Press.

Roth, W.-M. (2011). *Passibility: At the limits of the constructivist metaphor.* New York, NY: Springer.

Roth, W.-M. (2013). Toward a post-constructivist ethics in/of teaching and learning. *Pedagogies, 8*(2), 103–125.

Roth, M.-W. (2018). Translation and its consequences in qualitative social research: On distinguishing "the social" from "the societal." *FQS, 19*(1), 1–20. http://dx.doi.org/10.17169/fqs-19.1.2988

Roth, M.-W. (2020). *Adventures of mind and mathematics*. Cham, Switzerland: Springer.

Roth, W.-M., & Radford, L. (2011). *A cultural historical perspective on teaching and learning*. Rotterdam, the Netherlands: Sense.

Rubinstein, S. L. (1983). *Problemas de psicología general* [*Problems of general psychology*] (J. M. Güell, Trans.). Mexico DF: Grijalbo.

Rugg, H., & Shumaker, A. (1969). *The child-centered school*. New York, NY: Arno Press & The New York Times. (Original work published 1928)

Russon, J. (2017). *Sites of exposure: A philosophy essay on art, politics and the nature of experience*. Bloomington, IN: Indiana University Press.

Sachs, J. (2015). Aristotle: Motion and its place in nature. In *The internet encyclopedia of philosophy*. Retrieved April 9, 2015, from http://www.iep.utm.edu/

Sahlins, M. (1976). *Culture and practical reason*. Chicago, IL: The University of Chicago Press.

Said, E. (1994). *Culture and imperialism*. New York, NY: Vintage Books.

Salinas, U., & Miranda, I. (2018). Relating computational Cartesian graphs to a real motion: An analysis of high school students' activity. In N. Presmeg, L. Radford, M.-W. Roth, & G. Kadunz (Eds.), *Signs of signification. Semiotics in mathematics education research* (pp. 55–71). Cham, Switzerland: Springer.

Saxe, G. (2012). *Cultural development of mathematical ideas: Papua New Guinea studies*. Cambridge, England: Cambridge University Press.

Sayers, S. (2011). *Marx and alienation. Essays on Hegelian themes*. Hampshire, England: Palgrave Macmillan.

Segall, M. H., Campbell, D. T., & Herskovits, M. J. (1966). *The influence of culture on visual perception*. Indianapolis, IN: Bobbs-Merrill.

Scheiner, T. (2020). Dealing with opposing theoretical perspectives: Knowledge in structures or knowledge in pieces? *Educational Studies in Mathematics, 104*(1), 127–145.

Scheiner, T., & Pinto, M. (2019). Emerging perspectives in mathematical cognition: Contextualizing, complementizing, and complexifying. *Educational Studies in Mathematics, 101*(3), 57–372.

Schmandt-Besserat, D. (1992). *Before writing: From counting to cuneiform*. Austin, TX: University of Texas Press.

Scott, C. E. (1990). *The question of ethics. Nietzche, Foucault, Heidegger*. Bloomington/Indianapolis: Indiana University Press.

Sève, L. (2008). *Penser avec Marx aujourd'hui* [*Thinking with Marx today*] (Vol. 2). Paris, France: La Dispute.

Sfard, A. (2001). There is more to discourse than meets the ears: Looking at thinking as communicating to learn more about mathematical learning. *Educational Studies in Mathematics, 46*, 13–57.

Sfard, A. (2008). *Thinking as communicating*. Cambridge, England: Cambridge University Press.

Shapin, S. (1994). *A social history of truth. Civility and science in seventeenth-century England*. Chicago, IL: The University of Chicago Press.

Sheets-Johnstone, M. (2011). *The primacy of movement*. Amsterdam, the Netherlands: John Benjamins.

Silva D'Ambrosio, B., & Espasandin Lopes, C. (2015). Insubordinação criativa: um convite à reinvenção do educador matemático [Creative insubordination: An invitation to reinvent the mathematics educator]. *Bolema, 29*(51), 1–17.

Simon, R. I., & Dippo, D. (1986). On critical ethnographic work. *Anthropology & Education Quarterly, 17*, 193–203.

Spinoza, B. (1989). *Ethics including the improvement of the understanding* (R. Elwes, Trans.). Buffalo, NY: Prometheus. (Original work published 1667)

Stetsenko, A. (2017). *The transformative mind: Expanding on Vygotsky's approach to development and education*. New York, NY: Cambridge University Press.

Stetsenko A. (2020). Critical challenges in cultural-historical activity theory: The urgency of agency. *Kul'turno-istoricheskaya psikhologiya = Cultural-Historical Psychology, 16*(2), 5–18.

Stetsenko, A., & Arievitch, I. (1997). Constructing and deconstructing the self: Comparing post-Vygotskian and discourse-based versions of social constructivism. *Mind, Culture and Activity, 4*(3), 159–172.

Stocking Jr., G. W. (1966). Franz Boas and the culture concept in historical perspective. *American Anthropologist, New Series, 68*(4), 867–882.

Stocking Jr., G. W. (1982). *Race, culture, and evolution. Essays in the history of anthropology*. Chicago, IL: University of Chicago Press.

Tappan, M. (1998). Sociocultural psychology and caring pedagogy: Exploring Vygotsky's "hidden curriculum." *Educational Psychologist, 33*(1), 23–33.

Tarán, L. (1969). *Asclepius of Tralles: Commentary to Nichomacus' Introduction to arithmetic* (Vol. 59, part 4). Transactions of the American Philosophical Society.

Taylor, C. (1989). *Sources of the self*. Cambridge, MA: Harvard University Press.

Tihanov, G. (1998). Vološinov, ideology, and language: The birth of Marxist sociology from the spirit of lebensphilosophie. *The South Atlantic Quarterly, 97*(3–4), 599–621.

Toassa, G. (2014). Creating a materialistic psychology – sources and influence of Spinoza in Vygotsky's works. *International Journal of Liberal Arts and Social Science, 2*(5), 83–94.

Todorov, T. (1984). *The conquest of America. The question of the other.* New York, NY: Harper Perennial.

Todorov, T. (2000). *Éloge de l'individu* [*In praise of the individual*]. Paris, France: Adam Biro.

Tulviste, P. (1991). *The cultural-historical development of verbal thinking.* New York, NY: Nova Science.

Turner, V. (1988). *The anthropology of performance.* New York, NY: PAJ.

Underhill, J. W. (2009). *Humboldt, worldview and language.* Edinburgh, Scotland: Edinburgh University Press.

Vadeboncoeur, J. (2017). *Vygotsky and the promise of public education.* New York, NY: Peter Lang.

Valero, M. (2004). Postmodernism as an attitude of critique to dominant mathematics education research. In P. Walshaw (Ed.), *Mathematics education within the postmodern* (pp. 35–54). Greenwich, CT: Information Age.

Valero, P. (2009, January 28–February 1). Mathematics education as a network of social practices. In *Proceedings of the 6th Conference of European Research in Mathematics Education* (CERME 6). Lyon, France. Retrieved December 10, 2010, from http://www.inrp.fr/publications/edition-electronique/cerme6/plenary2-valero.pdf

Valero, P., & Knijnik, G. (2015). Governing the modern, neoliberal child through ICT research in mathematics education. *For the Learning of Mathematics, 35*(2), 34–39.

van der Veer, R., & Valsiner, J. (1994). *The Vygotsky reader.* Oxford, England: Blackwell.

Ver Eecke, P. (1959). *Diopante d'Alexandrie. Les six livres arithmetiques et le livre des nombres polygones* [*Diophantus of Alexandria: The six books on arithmetic and the book on polygonal numbers*]. Liège, Belgium: Desclée de Brower. (Original work published 1926)

Veresov, N. (1999). *Undiscovered Vygotsky. Etudes on the pre-history of cultural-historical psychology.* Frankfurt, Germany: Peter Lang.

Veresov, N. (2017). The concept of perezhivanie in cultural-historical theory: Content and contexts. In M. Fleer, F. González Rey, & N. Veresov (Eds.), *Perezhivanie, emotions and subjectivity: Advancing Vygotsky's legacy* (pp. 47–70). Singapore: Springer Nature Singapore.

Vernant. (1993). *L'homme Grec* [*The Greeks*]. Paris, France: Éditions du seuil.

Viète, F. (1983). *The analytic art* (T. R. Witmer, Trans.). New York, NY: Dover.

Vivier, L. (2020). Portée et usage du travail mathématique dans le cadre de la théorie des ETM. In M. Flores González, A. Kuzniak, A. Nechache, & L. Vivier (Eds.), *Cahiers du laboratoire de didactique André Revuz, n°21* (pp. 57–70). Paris, France: Université Paris Diderot.

Vološinov, V. N. (1973). *Marxism and the philosophy of language.* Cambridge, MA: Harvard University Press.

von Glasersfeld, E. (1995). *Radical constructivism: A way of knowing and learning*. London, England: Falmer Press.

von Humboldt, W. (1998). *On language: The diversity of human language-structure and its influence on the mental development of mankind*. Cambridge, England: Cambridge University Press.

Vygotsky, L. S. (1929). The problem of the cultural development of the child. *Journal of Genetic Psychology, 36*, 415–434.

Vygotsky, L. S. (1978). *Mind in society*. Cambridge, MA: Harvard University Press.

Vygotsky, L. S. (1979). Consciousness as a problem in the psychology of behavior. *Soviet Psychology, 17*(4), 3–35.

Vygotsky, L. S. (1987). *Collected works* (Vol. 1, R. Rieber & A. Carton, Eds.). New York, NY: Plenum Press.

Vygotsky, L. S. (1993). *Collected works* (Vol. 2, R. Rieber & A. Carton, Eds.). New York, NY: Plenum Press.

Vygotsky, L. S. (1994). The problem of the environment. In R. van der Veer & J. Valsiner (Eds.), *The Vygotsky reader* (pp. 347–348). Oxford: Blackwell.

Vygotsky, L. S. (1997a). *Collected works* (Vol. 4, R. Rieber, Ed.). New York, NY: Plenum Press.

Vygotsky, L. S. (1997b). *Educational psychology*. Boca Raton, FL: St. Lucie Press. (Original work published 1926)

Vygotsky, L. S. (1997c). *Collected works* (Vol. 3, R. Rieber & J. Wollock, Eds.). Problems of the theory and history of psychology. New York, NY: Plenum Press.

Vygotsky, L. S. (1998). *Collected works* (Vol. 5, R. Rieber, Ed.). New York, NY: Plenum Press.

Vygotsky, L. S. (1999). *Collected works* (Vol. 6, R. Rieber, Ed.). New York, NY: Plenum Press.

Vygotsky, L. (2004). Imagination and creativity in childhood. *Journal of Russian and East European Psychology, 42*(1), 7–97.

Vygotsky, L. S., & Luria, A. (1994). Tool and symbol in child development. In R. V. D. Veer & J. Valsiner (Eds.), *The Vygotsky reader* (pp. 99–174). Oxford, England: Blackwell.

Walkerdine, V. (1997). Redefining the subject in situated cognition theory. In D. Kirshner & J. A. Whitson (Eds.), *Situated cognition* (pp. 57–70). Mahwah, NJ: Lawrence Erlbaum.

Walshaw, M. (2013). Post-structuralism and ethical practical action: Issues of identity and power. *Journal for Research in Mathematics Education, 44*(1), 100–118.

Wartofsky, M. (1979). *Models, representation and the scientific understanding*. Dordrecht, the Netherlands: D. Reidel.

Wartofsky, M. (1984). The paradox of painting: Pictorial representation and the dimensionality of visual space. *Social Research, 5*(4), 863–883.

Watson, A., & Ohtani, M. (2015). *Task design in mathematics education. An ICMI study 22*. Cham, Switzerland: Springer.

Will, R. (2004). *The characteristic symphony in the age of Haydn and Beethoven*. Cambridge, England: Cambridge University Press.

Williams, J. (2016). Alienation in mathematics education: Critique and development of neo-Vygotskian perspectives. *Educational Studies in Mathematics, 92*, 59–73.

Williams, R. (1953). The idea of culture. *Essays in Criticism, 3*(3), 239–266.

Williams, R. (2005). *Culture and materialism*. London, England: Verso.

Wood, C. (1991). Introduction. In E. Panofsky (Ed.), *Perspective as symbolic form* (pp. 7–24). New York, NY: Zone Books.

Worsley, P. M. (1956). The kinship system of the Tallensi: A revaluation. *The Journal of the Royal Anthropological Institute of Great Britain and Ireland, 86*(1), 37–75.

You, H. (1994). Defining rhythm: Aspects of an anthropology of rhythm. *Culture, Medicine and Psychiatry, 18*, 361–384.

Zaporozhets, A. V. (2002a). Problems in the psychology of activity. *Journal of Russian and East European Psychology, 40*(4), 47–52.

Zaporozhets, A. V. (2002b). The development of sensations and perceptions in early and preschool childhood. *Journal of Russian and East European Psychology, 40*(3), 22–34.

Zavershneva, E. (2010). The way of freedom (On the publication of documents from the family archive of Lev Vygotsky). *Journal of Russian and East European Psychology, 48*(1), 61–90.

Zelnick-Abramovitz, R. (2009). Freed slaves, their status and state control in Ancient Greece. *European Review of History—Revue européenne d'histoire, 16*(3), 303–318.

Zevenbergen, R. (1996). Constructivism as a liberal bourgeois discourse. *Educational Studies in Mathematics, 31*, 95–113.

Index

abstract/concrete 1, 47, 48, 112, 205
activity 28, 29, 30, 50, 51, 110, 129
actuality 41, 54, 76
aesthetic 24, 25, 28, 29, 31, 71, 103, 126, 163
agency 69, 157, 185, 186, 204
agentic nature 189
agentic space 185–187, 201, 202
aktivität 29, 33, 59
al-gabr 193
algebra 44, 105, 129, 193, 224
alienation 206, 220, 221, 223
al-muquabala 193, 196, 197, 199
alterity 31, 213, 214, 217–219
ancient greeks 7, 8, 159, 186
antagonistic 179
apprenticeship 63
Arendt, H. 39, 208
Aristotle 7, 41, 42, 136, 168, 169, 174, 176, 181, 182, 208, 217
artefacts 26, 28, 52, 55, 87, 100
Arzarello, F. 19, 32, 122
assimilation 18
Atweh, B. 17, 19, 207
authenticity 79
authority 199, 201, 207, 209, 225, 226
autonomy 10, 15, 206, 227

Babylonian mathematics 53, 159, 162, 169, 181
Bacon, F. 7
Bakhtin, M. 18, 19, 24, 25, 37
Bartolini Bussi, M. 17, 19
becoming 1, 16, 24, 82, 191
Bednarz, N. 44
being 190, 212, 214–217, 221, 222, 225, 229
Bishop, A. 18, 19
Blunden, A. 61
Boas, F. 159, 160, 178, 181
body 6, 29, 32–34, 52, 55, 78, 84, 94, 95, 100, 104, 108, 109, 113, 128, 129, 136, 160, 176, 211, 213, 222–224
Boero, P. 19
Brousseau, G. 9, 11–13, 15, 16, 61, 211
Brown, T. 22

Cantoral, R. 17
care for others 225
carnival 25
categorical imperative 208, 227
child-centred pedagogy 20, 84, 92
chimpanzees 42, 43, 50, 56, 73, 74, 78
citizens in ancient Greece 168, 169, 171, 174, 177, 182
Cobb, P. 9–11, 14, 15, 33, 73, 184
commitment 224, 225, 227
common work 23, 24, 30, 33, 34, 52, 99, 135, 224, 225
communitarian ethics 219, 224, 228, 229
concept 86, 102, 103, 106
conceptual and contextual unity 133, 149
conceptual complexity 133, 134, 149
conflict 156, 158, 164, 165, 179, 229
consciousness 22, 28, 67, 71, 76, 79–84, 96, 99, 104, 164, 180
constructivism 10, 14, 24, 185, 227
contract 157, 210, 227
contradiction 93, 179, 222
criticism 198
cultural differences 156, 158–160, 178, 180
cultural ways of counting 53
culture 41, 43, 68, 73, 109, 116, 128, 156, 158, 160, 178

D'Ambrosio, U. 18, 207, 219
D'Amore, B. 59
Davydov, V. V. 89, 90, 103
deictics 122, 124
deyatel'nost' 29, 36, 59
dialectical materialism 1, 15, 25, 26, 34, 47, 157, 158, 166, 167, 175, 178, 185
didactic contract 13, 211
difference 25, 31, 40, 48, 70, 72, 106, 135, 148, 159, 179, 190, 191

Eagleton, T. 109, 157, 158, 162, 164, 167, 178, 180
educational reform 10
Elkonin, D. B. 68, 165
emancipation 190, 206
embodiment 1, 17, 26, 31–33, 40, 104, 108, 109, 111, 128, 129
empathy 196
empiricism 7, 108

Engeström, Y. 37
enlightenment 15, 31, 70, 72, 157, 189, 201, 227
Ernest, P. 10, 207
espace de travail mathématique 16
ethical commitment 35
ethics 1, 35, 36, 205–208, 210–216, 218, 219, 222, 224–229
Euclid 167, 172–174, 176, 177, 179–181
existence 212, 222, 227
expressivity 25, 78, 84, 220

Feuerbach, L. 32, 48, 69, 70, 110–112, 165, 190
flesh 212, 213, 215
forms of human collaboration 34, 35, 132, 133, 136, 206, 207, 225, 226
forms of knowledge production 34, 36, 135, 206, 226
freedom 157, 169, 185, 209, 227
Freire, P. 24, 35, 190, 206, 218, 220, 229
fulfillment 23
fundamental situation 12

Galperin, P. I. 67, 83
gegenstand 76–79, 82, 92, 106, 110, 111, 116, 131
generalisation 129, 130
genetic laws of cultural development 27, 65, 68
geometer practitioners 168
gestures 6, 33, 55, 109, 122, 126, 130
Godino, J. 22
Greek mathematical thinking 157
Greek practical numeric tradition 172

Hegel, G. 15, 23, 42, 50, 57, 58, 71–73, 76, 77, 84, 85, 130, 229
Heidegger 7, 103, 106
higher psychological functions 66, 69, 70
Hobbes, T. 205, 208, 210, 211, 227
homo economicus 161, 162, 178
homo symbolicus 161, 162, 178
hope 229
human dignity 217, 227
human labour 32, 33, 42, 48

idealism 67, 109, 111, 112
Ilyenkov, I. 15, 46, 47, 51, 84, 109, 165
imagination 122, 228
imitation 82
inferentialism 16
institutional knowledge 22
instrumentalist turn 27, 70
interaction 6, 19, 22, 34, 39, 62, 70, 88, 94, 133, 135, 166, 194, 203, 206, 220, 225, 229
internal-external dichotomy 68
internalisation 27, 64–70, 165
interpretation 8, 11, 25, 63, 135, 136, 138, 139, 149, 200
intersubjectivity 18, 19, 26, 39

joint labour 20, 25, 29–31, 33–35, 52, 53, 81, 82, 86, 92, 95, 96, 98, 99, 102, 121, 150, 201, 219, 220, 221, 223–225, 228, 229

Kant, I. 15, 39, 73, 76, 77, 183, 208, 210, 211, 227
Kieran, C. 44
knowing 49, 54, 82, 106
knowledge 21, 46, 49, 55, 73, 82, 219
knowledge as a generative capacity 40, 42, 49, 52, 53, 71, 73, 74, 78, 190
knowledge legitimation 207, 227

language 24, 25, 29, 37, 55, 130
legitimate peripheral participation 64
Leontev, A. N. 15, 28, 30, 80, 89, 109, 164, 221, 222
Lerman, S. 18, 19, 21, 39
Lévinas, E. 190, 191, 205, 208, 211–217, 225, 227–230

Mariotti, A. 17, 19
Marx, K. 31, 32, 47, 48, 58, 59, 69, 70, 84, 110, 111, 112, 113, 119, 131, 163, 164, 165, 189, 190, 218, 220, 221, 222, 225, 229
materialism 111–113
mathematics education 9, 36, 156, 228
matter 29, 32, 108, 129, 165, 213
meaning 70, 110, 158, 163
mediation 19, 23, 27, 33, 51, 53, 54, 58, 73, 86, 91, 128, 162, 205, 217, 218, 228
method 6, 8, 66
methodology 2, 9, 11, 13, 158
Morgan, C. 20
multiculturalism 180
multilingualism 180
multimodal 122
music 24, 29, 38, 57, 58, 62, 99, 166, 224

negation 61, 179
noticing 6, 75, 78, 79, 81, 84, 87, 89, 92, 101, 114, 117, 123

nut-cracking 42, 43

objectare 77
objectification 71, 83
objectification as an encounter 71, 73, 80, 83, 101, 105, 110, 129, 134
objectification processes 31, 34, 35, 76, 78–80, 82, 83, 85, 86, 92, 96, 98–102, 104–106, 110, 118, 119, 125, 198, 200, 202
Objekt 76–79, 106, 111, 116
ontogenesis 19
ontological 10, 25, 29, 31, 46, 49, 59, 77, 158, 189, 190, 214, 220, 222
ontology 214, 217
other 17, 18, 35, 81, 135, 205, 211–215, 217, 218, 223, 225, 227, 228
otherness 31, 99, 209, 225
overdetermination 165, 179

Pais, A. 22, 202
participation 82
pattern generalization 2, 44
perception 23, 33, 52, 66, 76, 81, 100, 105, 113–119, 121, 122, 128, 129, 159, 165
perezhivanie 69
phantasy 204
phylogenesis 19, 48, 54, 116
Piaget, J. 7, 8, 10, 11, 15, 16, 39, 61, 64, 72, 184, 201
Piggy Bank episode 4, 13, 14, 30, 33, 52, 81, 86, 92, 100, 105, 133, 134, 222
Plato 174, 176, 177, 179, 185, 217
platonic forms 42, 43
PME 19, 156
poetic 31, 103, 126, 130, 224
polis 169–172, 175, 179, 185
political 15, 18, 36, 72, 85, 113, 156, 158, 162–167, 169, 170, 175, 177, 178, 186, 191, 202, 205, 217–219, 227–229
positioning 191, 196, 200
potentiality 41, 54, 74, 76
power 206, 209
praxis 111, 112, 190
Presmeg, N. 27, 29, 56, 70
progressivism 185

race 178
racism 158, 159
rationality 16, 72, 160, 192, 200, 203, 209
Ratner, C. 156, 204
reality 10, 25, 37, 80, 110–113, 129, 180, 204
reason 217, 227
reification 77, 83
research questions 2, 6, 8, 9
resistance 158
responsibility 212, 213, 222, 223
rhythm 100, 126, 127, 130
Roth, W.-M. 21, 27, 29, 39, 60, 70, 89, 114, 207, 221
Rubinstein, S. L. 15, 226

Said, the 214–216
Saying, the 215, 216
scientific experiment 7
self 18, 72, 157, 184, 186–188, 190, 205, 209, 211, 227, 229
self as excess 189
self as lack 189
semiotic contraction 86, 102, 106, 122, 125, 128
semiotic means of objectification 86, 100, 106, 122, 126
semiotic node 86, 102, 106, 122
semiotic systems of cultural signification 186, 188, 201
sensibility 113, 129, 213, 215, 225
sensuous certainty 32, 48, 110
sensuous cognition 32, 33, 108, 109, 128
Sfard, A 16, 17, 20, 25
signs 25, 26, 27, 31, 40, 53, 54, 66, 215, 216
slavery 169
slaves 169–171, 179
solidarity 36, 59, 135, 137, 206, 215, 219, 222, 226, 229
speech 19, 30, 52, 122
Spinoza, B. 31, 67, 103, 189
student, concept of 20, 21, 23, 72
subject, concept of 71, 184, 185
subjectification processes 34, 35, 83, 183, 191, 192, 194, 202
subjectivities in the making 23, 34
subjectivity 23, 34, 36, 190, 191, 200, 212
submission 79, 207
subordination 79
subversion 61, 190

taken-as-shared 11, 33, 73
task design 132, 149

Tätigkeit 29, 37, 50, 59, 61, 112, 220, 221
teacher as a patriarch 23, 201
teacher, concept of 22, 23
tension 23, 39, 47, 165, 167, 176, 179, 186, 203, 229
theorematic mathematical thinking 179
theoretical principles 2, 7, 8, 9
theory in mathematics education 1, 2
theory of didactic situations 9, 11–16
thing-in-itself 77
Todorov, T. 18, 184
touching 110, 211, 223
transmissive teaching 21, 22, 24, 33, 39, 84, 88, 92, 203, 222
trust 223
truth 186, 188, 192, 198, 202

Valero, P. 22, 39, 202
Veresov, N. 67, 68, 69
voice 130, 191, 196, 200
Voloshinov, V. N. 24, 25
Von Glasersfeld, E. 10, 11, 15
vulnerability 212, 213, 215, 223, 225
vulnerable 222
Vygotsky, L. S. 9, 15, 16, 19, 20–22, 26–28, 30, 35, 39, 40, 49, 50, 61, 63–71, 80, 85, 103, 109, 119, 164, 180, 189, 229

Weltanschauung 6
Williams, J. 21
Williams, R. 158, 162–164, 178

Zaporozhets, A. V. 68, 121

www.ingramcontent.com/pod-product-compliance
Lightning Source LLC
Chambersburg PA
CBHW071432230926
51846CB00001B/206